Palgrave Studies in Media and Environmental Communication

Drawing on both leading and emerging scholars of environmental communication, the Palgrave Studies in Media and Environmental Communication Series features books on the key roles of media and communication processes in relation to a broad range of global as well as national/local environmental issues, crises and disasters. Characteristic of the cross-disciplinary nature of environmental communication, the books showcase a broad variety of theories, methods and perspectives for the study of media and communication processes regarding the environment. Common to these is the endeavour to describe, analyse, understand and explain the centrality of media and communication processes to public and political action on the environment.

Anne Hemkendreis • Anna-Sophie Jürgens
Editors

Communicating Ice through Popular Art and Aesthetics

palgrave
macmillan

Editors
Anne Hemkendreis
University of Freiburg
Freiburg im Breisgau, Germany

Anna-Sophie Jürgens
Australian National University
Canberra, ACT, Australia

Alfried Krupp Wissenschaftskolleg, Greifswald
SFB 948 „Heroes – Heroizations – Heroisms", University of Freiburg
Australian National Centre for the Public Awareness of Science,
The Australian National University

ISSN 2634-6451 ISSN 2634-646X (electronic)
Palgrave Studies in Media and Environmental Communication
ISBN 978-3-031-39786-8 ISBN 978-3-031-39787-5 (eBook)
https://doi.org/10.1007/978-3-031-39787-5

Konrad Lenz (photographer, cover image)

This Palgrave Macmillan imprint is published by the registered company Springer Nature Switzerland AG.
The registered company address is: Gewerbestrasse 11, 6330 Cham, Switzerland

Paper in this product is recyclable.

Foreword

Drawing on a range of perspectives, this splendid collection is at the leading edge of a significant turn in the arts, humanities and social sciences that insists on the importance of the use of ice as an active aesthetic resource and tool within contemporary and future artwork, performance, media and popular culture to address the reality and severity of the climate crisis as well as ways to reimagine our world anew. What makes this contribution especially important is the way it not only further scrutinises these stories and experiences about ice through art and popular media practices but breaks with settler-colonial fantasies and frontier masculinities and its aesthetic responses to think more creatively about the centrality of ice in the future of human and non-human life.

Ice and particularly melting ice might be a source of dread for most of us who will be impacted by rising sea levels, but here in the collection there is a counter to dread that offers compelling ways to expand our political imagination about how we tell stories about ice and our relationship to ice in ways that could help us reimagine and remake the planet in ways more liveable to all. This collection gives us plenty of fascinating ways to delve into thinking and acting with ice and invites new ways of living with ice to make us question what kind of world we collectively want to live in now and in the future.

Lecturer in the Media Studies Department
and a research scholar in the Department of
Gender and Women's Studies
University of California, Berkeley
Berkeley, CA, USA

Lisa E. Bloom

Acknowledgements

Extreme cold, sublime glow and tremendous soundscapes make up the sensual dimensions of ice and have equally inspired science, art and technology since the nineteenth century. They generated cosmological models used to explain the world, such as Hanns Hörbiger's 1913 *Welteislehre* ("World Ice Theory"), which was largely inspired by the conquest of the North Pole. These entered the history of science and popular culture through a sophisticated, broadly effective communication that reached experts and non-experts alike. To this day, events such as the "Antarctic Biennale" of 2017—which brought together a hundred "artists, scientists and visionaries" on an expedition ship—bear witness to the close intertwining of Polar regions, environmental research and arts.

Future knowledge and design—following the current focus of "ice research" on climate-related issues—are based on physical experience and sensual communication to an audience. There is a fine line between objective-knowledge generation and emotional involvement for the purpose of communicating climate awareness. This tension has its roots in the nineteenth century. Inspired by the search for the lost "John Franklin Expedition" (1845), Arctic panoramas, dioramas and theatre plays were created in Europe and the US as spectacular productions in which science communication and entertainment were inseparably linked. Technical innovations—such as dramatic lighting, to suggest an aurora borealis—caused the audience to be sensually affected by icy immersion effects that evoked direct participation in the Polar expeditions. To this day, ice in environmental art and science communication serves as a strategy for a hybrid mode of reception that oscillates between emotional and epistemological participation. The communication of urgency—to protect our environments more actively and prevent further exploitation of nature—goes hand in hand with attempts to generate and stage aesthetical evidence.

The proximity of art and science, its historical roots and future potentials are the focus of this edited volume and the three preceding international online conferences of 2021—entitled *Ice (St)Ages* 1–3. We would like to take this opportunity to wholeheartedly thank our supporters, colleagues, student collaborators (in particular Rishika Nair Prabhakaran) and friends who have made these interdisciplinary ice explorations and their publication possible: the Collaborative Research Centre (SFB) 948 "Heroes—Heroisations—Heroismen" at the University of Freiburg (Germany); the Alfried Krupp Wissenschaftskolleg, Greifswald (Germany); and the Australian National Centre for the Public Awareness of Science of the Australian National University (Australia). This volume has benefited immensely from the invaluable critical comments, intellectual input, feedback and expert editorial support of Dr Rebecca Hendershott, as well as the highly professional support of Kristina Sieling and Karina Judd.

Of course, we are also indebted to the extraordinary speakers who have contributed to the interdisciplinary ice adventures and intellectual enjoyment of our *Ice (St)Ages* conferences, many of whom are among the authors of this book. All of our international authors—humanities scholars, scientists, artists, science communicators and social scientists—have worked constructively and collaboratively to prepare their texts for this publication, for which we are sincerely grateful. In this book we cherish and celebrate their different perspectives, disciplines and backgrounds, which is why all chapters have their own individuality while staying focused on the cultural value of ice.

The richness of the scholarly perspectives gathered in this volume invites us to reflect on the different kinds of ice knowledge that can inform a Western understanding of nature and lead to a healthier relationship between humans and the non-human world. Our perception of, and dealing with, ice can only change if our conception of the Polar regions, and the cryosphere in general, is examined for its ignoramuses and invisible, historical and current power structures. When ice appears as an active agent in aesthetic contexts of knowledge transfer, processes of ecological knowledge generation and moments of encounter between nature and humans are stimulated. Even more, ice—as a changeable element—makes it possible to think about alternative, that is more sustainable, ways of living.

Ultimately, despite ice being such a cold topic, we warmly thank the team at Palgrave Macmillan, the anonymous reviewers and our colleagues and friends who have enthusiastically endorsed this adventurous project that bridges Science, Science Communication, Art and Popular Culture.

The Editors
Canberra and Freiburg, December 2022

Praise for *Communicating Ice Through Popular Art and Aesthetics*

"A vibrant, very contemporary contribution to the rising field of 'Ice Humanities,' intersecting art with science, film and music with ethnography, and geographies with planetary humanities this volume truly stages and communicates the vastest, vulnerable, and whitest species on Earth."

—Sverker Sörlin is Professor in the *KTH Environmental Humanities Laboratory* in Stockholm (Sweden) and the co-editor of Ice Humanities (Manchester UP 2022)

"It matters what stories tell stories, what visions create visions. In order to change a story, we have to know what it is. Only then can we open ourselves to new narratives, new visions. This book manages to do both. It analyses how our imagination of ice is shaped by masculine and colonial fantasies. But it also demonstrates how a novel 'ice science' may emerge from removing the traditional boundaries that separate science, art and humanities. With its well-written analyses, new insights and, not least, uplifting visions of new approaches, collaborations and opportunities, this book is a joy to read."

—Kirsten Thisted is an Associate Professor for the Minority Studies Section at the Department of Cross-Cultural and Regional Studies at the *University of Copenhagen* in Denmark and an adjunct professor at the Department of Culture, Language and History in Ilisimatusarfik at the *University of Greenland*

"This original, interdisciplinary volume examines how we engage with melting ice from a variety of perspectives, bringing into dialogue views from ecology, oceanography, geography, geopolitics, science communication, comparative literature, art history, media studies, musicology, popular entertainment studies and more. This book makes a compelling attempt to understand ice not as an object but as an agent with which we must interact differently. The contributions thus convey a fascination with ice that transcends the traditional Arctic sublime and indicates new directions for a multidisciplinary (Ant)Arctic discourse. Ultimately, the highly recommended volume points to the extreme urgency of political action to curb global warming."

—Evi Zemanek is Professor for Media-Cultural-Studies and Media Ecology at the *University of Freiburg* in Germany

"Our planet's glaciers are melting. Ice loss from the Greenland and Antarctic ice caps is accelerating, and sea level rise, which is already threatening coastal communities, will disrupt or displace millions of people this century. Actions we take—or fail to take—in response to these threats will impact our planet for centuries. But how do we talk about something so big? How do we make people notice, care, act? Scientists and policymakers can't respond to climate change and sea level rise without an engaged public committed to change. In this exciting collection, scholars from around the globe explore how artists, activists, scientists, and communicators are responding to the climate emergency by creating narratives and images that demand attention, create a sense of urgency, and even give us cause to smile, wonder—and hope."

—Rebecca Priestley, Professor of Science in Society, *Te Herenga Waka—Victoria University of Wellington* Author of *Fifteen Million Years in Antarctica and Dispatches from Continent Seven*

Contents

Notes on Contributors

Dana M. Bergstrom is an applied, integrative Antarctic ecologist and author. Her principal position since 2002 has been at the Australian Antarctic Division. She holds a visiting professorship in the Faculty of Science at the University of Johannesburg, South Africa, and a visiting fellowship at the University of Wollongong, Australia. She specialises in turning disparate data into coherent science and knowledge.

Frédéric Bouchard is Assistant Professor of Climate Change in the Department of Applied Geomatics at the University of Sherbrooke, Canada, and is also an affiliated researcher at Geosciences Paris-Saclay (GEOPS), Orsay, France. First trained as a geologist, he later specialised in physical geography (geomorphology, climatology) and water sciences (sedimentology, paleolimnology), specifically in high-latitude regions. His research projects explore the multiple impacts of permafrost degradation on aquatic ecosystems and northern communities across the circumpolar Arctic (northern Canada, Siberia). Bouchard is a founding member of the "Frozen-Ground Cartoons" project and is deeply involved in scientific outreach, communication and education about permafrost and Arctic research.

Stefan Buchenberger works in the Department of Cross-Cultural Studies at the Kanagawa University, Yokohama, Japan. He holds a PhD in Japanese Studies from the Ludwig Maximilian University, Munich, Germany, in 2004. He is involved in the study of graphic narratives at the International Comparative Literature Association (ICLA), where he is the co-founder and co-chair of the standing Research Committee on Comics

Studies and Graphic Narrative. He is also a member of the Japanese Comparative Literature Association and the Japanese Germanics Society. His other areas of research are mystery fiction, horror movies, intertextuality and orientalism.

Klaus Dodds is Professor of Geopolitics and executive dean of the School of Life Sciences and Environment at Royal Holloway, University of London. He is a fellow of the UK's Academy of Social Sciences and honorary fellow of British Antarctic Survey. He has written and edited several books on ice including a co-edited collection *Ice Humanities* (Manchester University Press 2022 with Sverker Sörlin) and *Ice: Nature and Culture* (Reaktion 2018).

Matteo Farinella received a PhD in Neuroscience from University College London in 2013. Since then he has been combining his scientific expertise with a life-long passion for drawing, producing educational comics, illustrations and animations. As a Columbia University Presidential Scholar in Society and Neuroscience, he has researched the role of comics in science communication and worked with scientific institutions all over the world to make science more clear and accessible. Since 2019 he has worked as a scientific multimedia producer at Columbia's Zuckerman Institute. His work can be found on matteofarinella.com.

Laurence Grove is an expert of French and text/image studies and director of the Stirling Maxwell Centre for the Study of Text/Image Cultures at the University of Glasgow, UK. His research focuses on historical aspects of text/image forms, in particular *bande dessinée*. He is the president of the International Bande Dessinée Society. As well as serving on the consultative committees of a number of journals, he is the co-editor of *European Comic Art* and makes frequent media appearances. Laurence (also known as Billy) has authored (in full, jointly or as editor) eleven books, including *Comics in French* (Berghahn 2010 and 2013) and approximately sixty chapters or articles. He co-curated the *Comic Invention* exhibition (The Hunterian, Glasgow 2016; Clydebank Museum 2017) and *Frank Quitely: The Art of Comics* (Kelvingrove, Glasgow 2017), and has long-term hopes of seeing a National Comics Centre for Scotland.

Anne Hemkendreis works as an academic researcher for the interdisciplinary project 948 "Heroes—Heroizations—Heroisms" at the Freiburg University, Germany. She is an associate senior lecturer at the Humanities Research Centre, Australian National University, Canberra, and a member of the Young Academy of the German National Academy of Sciences Leopoldina. Prior to this, Anne worked as a fellow at the Alfried Krupp Wissenschaftskolleg, Greifswald, and as a research assistant at the Leuphana University, Lüneburg (both in Germany). She has taught at various universities, including the University of the Arts in Berlin, while performing on stage as an artist in physical theatre and circus arts. Anne is interested in how female artists question heroic imaginaries in times of climate change with a special emphasis on the Polar regions and the cosmos. An additional passion of Anne is examining environmental circus shows as an artistic mode of communicating posthuman perspectives. She is the editor of this volume and a co-editor of the special issue *Climate Heroism* in helden, heroes, héros: E-Journal zu Kulturen des Heroischen (2022).

Karina Judd is a PhD candidate in the Australian National Centre for the Public Awareness of Science at the Australian National University, Canberra, Australia. Her current research is on inclusive science communication and practice around the world, building on her past professional experiences as a science communicator, science outreach facilitator, environmental geologist and geoarchaeologist. Karina is supported by the Australian Government Research Training Program.

Anna-Sophie Jürgens is a Senior Lecturer in Science Communication at the Australian National Centre for the Public Awareness of Science, Australian National University (ANU), and the head of the POPSICULE, ANU's Science in Popular Culture and Entertainment Hub. Her research explores the cultural meanings of science and science and comic performance in different media. Anna-Sophie has written on comic parasites and Joker viruses in (animated) fiction, environmental fragility in street art and clowns and scientists in comics and popular theatre in numerous academic journals. Her recent books include *Circus and the Avant-Gardes* (co-editor, Routledge 2022) and *Circus, Science and Technology: Dramatising Innovation* (editor, Palgrave Macmillan 2020). Anna-Sophie is the editor-in-chief of the peer-reviewed open access online journal *w/k—Between Science and Art* (English section) and associate editor of the *Journal of Science & Popular Culture*.

Nanna Katrine Lüders Kaalund is a postdoctoral research associate at Aarhus University, Denmark, where she holds a Carlsberg Fellowship for the project "Economizing Science and National Identities: The Royal Greenland Trading Department and the Making of Modern Denmark and

Greenland". She previously worked at the University of Cambridge and the University of Leeds, and is the author of the book *Explorations in the Icy North*, which was published with the University of Pittsburgh Press in 2021. Her research examines the intersection of Arctic exploration, race, print culture, science, religion and medicine in the modern period with a focus on the British, North-American and Danish imperial worlds.

Elizabeth Leane is Professor of English in the School of Humanities at the University of Tasmania, Australia. She holds degrees in physics and literary studies and her first monograph, *Reading Popular Physics* (2007), focused on communication of science. Since moving to the Polar "gateway" of Hobart, she has been investigating the stories we tell about Antarctica, how they inform our ideas and attitudes towards the continent and how they can enable new ways of thinking about our relationship to place and environment. Elle is the author of two other monographs—*South Pole: Nature and Culture* (2016) and *Antarctica in Fiction* (2012)—and the co-editor of five collections. She is Arts and Literature Editor of *The Polar Journal* and a former Australian Antarctic Arts Fellow.

Ben Nickl is a lecturer in the Department of Comparative Literature and Translation Studies at The University of Sydney, Australia, and serves as an editor of the *Global Germany* series and associate editor for the *Journal of Science & Popular Culture*. He is working on popular cultural technologies that shape and mass produce and influence our human feeling; among them currently are digital laughter and synthetic emotion in poetic and cinematic mass culture and TV mainstream entertainment. What can pop cultural analysis do to engage the big problems of our time, like climate change? How does mass culture mediation shape and give access to and allow us to intuit seemingly immediate yet curiously unknowable parts of the environment we live in? What kind of language and what tools does it provide for this? These are the questions Ben seeks to respond to in research works such as this chapter and forthcoming work on artificial affect in popular culture.

Hanne E. F. Nielsen is Lecturer in Antarctic Law and Governance at the Institute for Marine and Antarctic Studies, University of Tasmania, Australia. She specialises in representations of Antarctica and has a particular interest in the commercial history of the continent. Hanne serves on the steering committee of the Scientific Committee on Antarctic Research (SCAR) and Standing Committee on Humanities and Social Sciences; is a lead on the SCAR Tourism Action Group (Ant-TAG); and is a past president of the Association of Polar Early Career Scientists (APECS).

Gabby O'Connor is a Melbourne-born, NZ-based installation artist, educator and climate communicator. In 2015 and 2016, O'Connor spent time researching sea-ice crystals while collaborating with oceanographers. This research resulted in the exhibitions *Studio Antarctica* and *Data Days + Studio Antarctica* and more. More recently, O'Connor's work (including PhD research) focuses on communicating risk around climate change and sea level rise through large-scale public art projects exhibited throughout NZ. These exhibitions include a participatory element so that the public can contribute to the making of a contemporary artwork.

Carolyn Philpott is Senior Lecturer in Musicology at the University of Tasmania's Conservatorium of Music and an adjunct senior researcher at the Institute for Marine and Antarctic Studies. She has written widely in journals and books in the fields of musicology and Antarctic studies, including in *The Musical Quarterly*, *Musicology Australia*, *Popular Music*, *The Polar Journal* and *Polar Record*. She has also written a monograph on the music of Australian composer Malcolm Williamson (Lyrebird Press 2018) and a co-edited collection titled *Performing Ice* (Palgrave Macmillan 2020).

Johannes Riquet is Professor of English Literature at Tampere University, Finland. He is the author of *The Aesthetics of Island Space: Perception, Ideology, Geopoetics* (OUP 2019) and the co-editor of *Spatial Modernities: Geography, Narrative, Imaginaries* (Routledge 2018) as well as *Imaging Identity: Text, Mediality and Contemporary Visual Culture* (Palgrave Macmillan 2019). His research interests include spatiality, literary geography, visual culture, travel writing, diaspora and mobility. He is the principal investigator of the collaborative project "Mediated Arctic Geographies" (Academy of Finland, 2019–2023). He is also working on a monograph on interrupted railway journeys in fiction and visual culture.

Birgit Schneider is Professor of Knowledge Cultures and Media Environments in the Department of European Media Studies at the University of Potsdam, Germany. She studied art and media studies as well as media art and philosophy in Karlsruhe, London and Berlin. After initially working as a graphic designer, she worked from 2000 to 2007 in the research department "The Technical Image" at the Humboldt University, Berlin, where she received her doctorate. Since 2009, she has been researching in the context of fellowships in the European Media Studies Department at the University of Potsdam as well as in Munich, Weimar and Cambridge, UK. Her current research focuses are images and perceptions of nature, ecology and climate change, diagrams, data graphics and maps as well as images of ecology.

Ylva Sjöberg is Associate Professor of Geography in the Department of Geosciences and Natural Resource Management at the University of Copenhagen, Denmark. She is trained as a geographer and specialised in Arctic hydrology and permafrost during her PhD and postdoctoral studies at Stockholm University and the US Geological Survey in Alaska. Her research on the water cycle in Arctic permafrost environments leverages on both field-based and modelling techniques. Sjöberg is also a founding member of the "Frozen-Ground Cartoons" project and has been involved in art-science collaborations, including an audio-based project exploring permafrost from the perspective of a mosquito.

Craig Stevens is an oceanographer based in New Zealand with a joint position at the National Institute for Water and Atmospheric Research and as a professor at the University of Auckland. His research focus is on extreme ocean environments like Antarctic oceanography, tidal turbulence, marine heatwaves and greenhouse gas uptake in the Southern Ocean. He grew up in South Australia and studied in Australia and Canada before moving to New Zealand in the 1990s. He is a past president of the New Zealand Association of Scientists and has participated in 14 Antarctic field campaigns and nearly 50 ocean experiments.

Peta Tait is an academic at La Trobe University, Australia, a playwright and a fellow of the Australian Academy of the Humanities. She has written 70 scholarly articles and chapters and recent books include the authored: *Forms of Emotion: Human to Nonhuman in Drama, Theatre and Performance* (Routledge 2022), *Theory for Theatre Studies: Emotion* (Bloomsbury 2021), *Fighting Nature: Travelling Menageries, Animal Acts and War Shows* (Sydney University Press 2016) and *Wild and Dangerous Performances* (2012); the co-edited *Feminist Ecologies: Changing Environments in the Anthropocene* (2018) and *The Routledge Circus Studies Reader*; and the edited *The Great European Stage Directors volume one* (2018). Her current collaborative "Towards an Australian Ecological Theatre" project is funded by the Australian Research Council.

Stephanie von Spreter is Doctoral Research Fellow in Art History at University of Tromsø (UiT) – the Arctic University of Norway, member of the research group "Worlding Northern Art" (uit.no/research/wona) and the transdisciplinary research project "Urban Ecologies: City Sensing Beyond the Human" (urbanecologies.no). She recently wrote "Pia Arke and 'Arctic Hysteria': Visual Repatriation and the Problematics of a 'Lost' Artwork" (Kunst og Kultur 2022) and "Feminist Strategies for Changing the Story: Re-imagining Arctic Exploration Narratives Through (the Staging of) Photographs, Travel Writing and Found Objects" (*Journal of Aesthetics and Culture* 2021). She previously served as artistic director of Fotogalleriet, Oslo, and worked for the 3rd, 4th and 5th Berlin Biennale, the 50th Venice Biennale and Documenta11. She also works as an independent curator and writer.

Graham J. Walker is Lecturer in Science Communication at the Australian National Centre for the Public Awareness of Science, Australian National University, Australia. His current research and engagement investigates international capacity building, culture, climate change, psychological aspects and co-design in science communication, with a focus on science centres, science shows and hands-on workshops and using these methods to engage with social and environmental issues. He also teaches in these areas and creates, performs and delivers training on science shows globally. Twitter: @DrGrahams.

List of Figures

Here Be Science Show Dragons: Ice, Icons and Metaphoric Approaches to Climate Change Communication

Materiality of Time: Polar Ice as a Medium for Ecological Art for the Tempered Zones

Hard Ice, Soft Snow? Transnationalism, Spectatorship and the Arctic Sublime in *Chasing Ice* (2011) and *Silent Snow* (2011)

Frozen-Ground Cartoons—Revealing the Invisible Ice

On the Visual Narratives of Ice in Popular Culture: Comics on Ice, Icy Villains and Ice Science

Melt for Me: Communicating Ice Empathy Through the Plasticity of Disney

On the Aesthetic Facets of Ice Urgency: Some Final Reflections

Staging Ice and Ice Stages in Science, Science Communication and Aesthetic Experience

Communicating Loss: Ice Research, Popular Art and Aesthetics: Introduction

Anne Hemkendreis and Anna-Sophie Jürgens

Communicating With, and About, Ice

Water, in its various states of aggregation—such as ice—connects us humans with our environment. Thinking with, and about, ice forces reflection on the essence of nature-human relationships; not in a demarcation of nature and culture, but in their mutual interpenetration and dependence. In its frozen state and when it melts, ice has both a form-assuming and a form-dissolving quality; as a shape-shifter, it becomes a metaphor for the entanglements of humans with nature as a dynamic system that is deeply troubled. In this regard, the increasing loss of ice—at the polar regions and elsewhere—mirrors our own physical and immediate

A. Hemkendreis (✉)
University of Freiburg, Freiburg im Breisgau, Germany
e-mail: Anne.hemkendreis@sfb948.uni-freiburg.de

A.-S. Jürgens
Australian National University, Canberra, ACT, Australia
e-mail: anna-sophie.jurgens@anu.edu.au

A. Hemkendreis, A.-S. Jürgens (eds.), *Communicating Ice through Popular Art and Aesthetics*, Palgrave Studies in Media and Environmental Communication,
https://doi.org/10.1007/978-3-031-39787-5_1

endangerment by climate change. This is why ice is commonly used as a means of communicating the need to change our behaviour by fundamentally reflecting on our perception of, and attitude towards, nature in its current and historic dimensions.

In times of climate change, scientists are faced with the challenge of making their knowledge and data accessible to wider audiences and reaching different people through different avenues and media. This is why scientists turn to artists, to give the abstract contexts of climate change a sensual and tangible form—for example, by immersive effects or emotional address. But artistic work with ice is far more than an instrument of knowledge communication in a linear sense, that is, as a one-way street.

For example, Astrida Neimanis's observation of the wateriness of the human body underlines that humans are inseparable from "pressing ecological questions" (2019, 1). Thinking with, and about, water in its various states enables insight into the fluid boundaries between the human, the non-human and the more-than-human. The author uses a materialist philosophical approach to outline an alternative to the Anthropocene worldview that has put humans and our needs in its centre (Neimanis 2019, 2).

Countering an abstract and white European universalism, Neimanis proposes an encounter between humans and nature which fosters insights into our own physical connection to, and entanglement with, this planet. Our dependency on the surrounding world has fostered a development in the arts in which the artistic material (here ice) does not serve as a meta-reflection on the nature of art—rather, it communicates that in the end, everything is nature.

In Western culture, we also encounter ice in its different forms in various types of science-related visual media, including films, cartoons and science shows. The sensual quality of ice—its coldness, its glow and its changeability—makes it a popular medium of communication in a climate-related context. More so, ice (melt) "carries a potent material politics" (Randerson 2018, 137). Melting polar ice is causing sea levels to rise sharply, threatening islands and coastal regions in particular. Each reef, each lake, each glacier has produced its own set of scientific experts who are now faced with the challenge of bringing together the various knowledge gains and communicating them as an interwoven system (Randerson 2018, 138). Ice can generate this sense of entanglements when used in scientific and aesthetic discourses. A "vital task" for ice art, ice shows or ice films is "to stir up an affective response to the rapidly shifting ecological conditions" (Randerson 2018, 139). Personal relationships to climate change are brought about by an emotional encounter with ice as an active counterpart, evoking a range of effects, from overwhelm to identification and self-reflection. Ice art opens up a space of sensual experience in which ice appears as an active agent of knowledge production and reflection: Crucially, the reception of ice in art promotes reflection on one's own relationship to the environment, the history of one's own perception and one's own agency.

Environmental Activism and the Sublime

Our imagination of ice is shaped by masculine and colonial fantasies that have remained unquestioned until recently. In particular, the so-called eternal ice—as a polar phenomenon and our associated emotions—draws from the nineteenth-century colonial period, which focused on conquests of the North and South Pole. These events lead to numerous science-related publications on heroic deeds. The ice was generally imagined as an empty stage on which heroic deeds could be acted out, following the narrative of man-against-nature (Bloom 1993). The hidden violence within these imaginaries can be traced to modern day; particularly, the silencing of women and animals within the (Ant)Arctic journeys, the denial of knowledge from the Inuit for the so-called heroes' success and the local communities' ongoing suffering (Bloom 2022). Today, Ant(Arctic) tourism and the warming of the poles make it possible for almost everyone—who has profited from the wealth of Western societies—to step into the footsteps of former polar heroes and heroines, inflicting further hidden violence. As Elena Glasberg noted, "the more people track the poles, the more they empty the ice of its own materiality or liveliness" (Glasberg 2011, 222).

Today, the irreversible loss of ice, measured and conveyed by scientists and broadcasted in the news, has slowly led to a shift within the perception of ice and its cultural meaning. Understood as a natural archive, ice expresses cultural injustices and environmental changes, while addressing us emotionally and stressing the need to act. Artists react to this by giving voice to the ice itself and paying homage to its liveliness. They let the ice speak and communicate its hidden knowledge. This development led to a shoulder-to-shoulder of arts, science and activism.

Thus, in contrast to its increasing loss, ice in our everyday culture seems omnipresent and available. For example, in June 2016, musician Ludovico

Einaudi posted a video on his YouTube channel called "Elegy for the Arctic". In the video, Einaudi calls for the protection of glaciers by playing his sad composition on a grand piano on an artificially created, floating platform in front of a glacier in the Arctic Ocean (Fig. 1). The performance emerged from an initiative by Greenpeace and resulted in an online petition—seen by more than eight million subscribers—to save the Arctic (Einaudi 2016).

This Greenpeace campaign brought together activism, science communication and art. Einaudi travelled to the north aboard the Greenpeace ship Arctic Sunrise on the eve of a significant event: the meeting of the OSPAR Commission, an association under international law for the protection of the North Sea and the North-East Atlantic. In the video, he performs his elegy against the backdrop of the Wahlenbergbreen glacier in Svalbard, Norway. At the very moment of the elegy's climax, part of the glacier breaks off and falls into the water. The effect is fascinating and confusing at the same time. Is it the strong connection between the music and the powerful images of nature that briefly creates a miraculous unity

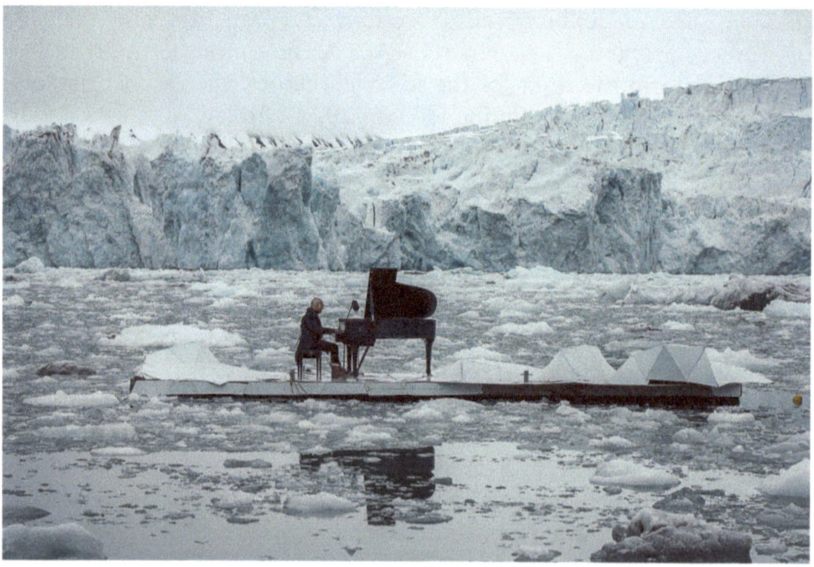

Fig. 1 Ludovico Einaudi (Pianist/Composer): Performance in the Arctic Ocean, 2016 © Pedro Armeste, Greenpeace

between humanity and the environment? Or is it the music (and thus culture) itself that leads to the catastrophic effect, hinting at the vulnerability of the ecosystem in the far North? This uneasiness makes us pause in awe and wonder.

Appreciating ice as an aesthetic means of sensual appeal, engagement and communication has its origins in the history and visual representation of natural disasters. The 1755 Lisbon earthquake is considered the first natural catastrophe whose images were widely disseminated in the media of the time (Weber 2018). In the visual discourse surrounding this event, documentary aspirations and aesthetic fiction unified in their goal for both representative effects and narrative structures (Scholz 2021). The visual circulation of this catastrophe caused consternation across Europe, and incited interest in aesthetic reception of all kinds of catastrophes, including ship disasters. One of the most popular and unique paintings of ship disasters in the Romantic period is Caspar David Friedrich's famous *Sea of Ice* (1823/24), which has been associated with the beginning of ecological thinking in art (Grawe 2001; Rautmann 1991). It is thus no coincidence that Friedrich's *Sea of Ice* became a reference for climate-critical activism and activist art of the 1980s and 1990s. Today, this famous painting is the iconographic role model for the majority of spectacular ice art—including Einaudi's performance, which positions the viewer on an (artificial) ice-floe alongside the artist.

The reference to Romantic aesthetics addresses how human subjects are physically affected by overwhelming, indescribable and awe-inspiring natural phenomena that dignify, excel or frighten us. The vulnerability of Einaudi's icy environment, however, turns this effect on its head and imparts a troubling message: not only are humans threatened by invisible forces made manifest (e.g. in breaking icebergs), but entire habitats are subject to drastic, threatening changes. In these contexts, nostalgia—a wistful, sentimental yearning for a past time, irretrievable state or irrecoverable condition—evokes the feeling, if not experience, of loss; in this case, the loss of the ice. By referring to, and evoking, these complex aesthetics, Einaudi's performance taps into a fascinating cultural reservoir of aesthetic and emotional experience: It is a conversation between the past, present and future. His performance is impressive and alarming, enjoyable and thought-provoking—and, by using art and aesthetics, it communicates a powerful environmental message about urgency and, ultimately, human responsibility. Using (artificial) ice as a platform to tell a story about the catastrophe of ice melt, Einaudi's enticing (video) performance is explicitly staged as a show for an international, remote audience. The soundscape created by the piano and the crashing iceberg builds a bridge between the viewers of the video (however distant they may be) and the Arctic by interweaving nature and culture. Einaudi's artistic comment on environmental change and exigency is conveyed through staging, setting and visual and audible effects. Artistically interacting with the breaking iceberg, Einaudi literally plays with sublime aesthetics and, more precisely, the environmental sublime. The sublime is a cultural concept and an artistic means of confronting a subject with boundless (natural) forces. As a threshold experience, the sublime mediates between the visible and the invisible, between the form and the formless—which is why ice, as a shape-shifter, is used for its evocation. The sublime is based in the philosophy of enlightenment by Immanuel Kant and Edmund Burke and means an overwhelming experience; it is now connected with the ungraspable horror of climate change (Horn and Bergthaller 2019). Artists inspired by Kant's and Burke's philosophy are attracted by the sublime as sensual experiences and use it to highlight humans' physical dependence on the planet (Morgan 2016, 4).

The sublime as a tool for communicating climate change has been hotly debated in relation to its roots, intensity and effect, especially when it comes to art. Some authors criticise evoking ecological awareness through sublime aesthetics (Brady 2013, 119). They point to the roots of the sublime in European Universalism: the idea of a white and male-orientated domestication of nature. This is why representing climate change by sublime aesthetics often risks overwriting ecological awareness with latent colonial fantasies of finality—the extinction of species, the end of Indigenous people and their knowledge (Plumwood 1993). As Val Plumwood argues, the rationality of 'Western' culture and its aesthetic offspring (such as the sublime) have been systematically unable to acknowledge humans' dependence on nature and those defined as inferior 'others'. Birgit Schneider similarly points to the connection of the sublime to the history of spectacle and political propaganda, emphasising the danger of sublime aesthetics in climate art to immobilise climate action and produce fake news (2021). In contrast, others defend the use of the sublime in polar art, stressing its potential to evoke feelings of connectedness (Boetzkes 2020, 40–41). As Amanda Boetzkes points out, (climate-) science-related art can use the sublime to yield a sense of the earth's excess and the limit of representational form (2010, 110). According to Boetzkes, the environmental sublime can communicate the abstract interrelations of climate change phenomena, and also lead to a conscious encounter between humans and the dynamic , e.g. invisible, forces of nature. A recognition of these forces allows nature to be imagined as something connected to humans as a counterpart. The potential of the sublime, which transcends the imagination, thus combines aesthetic experience and ethical significance.

In Einaudi's performance, the sublime arctic is more than a passive stage or scenic background. The ice shapes the music and interacts with the performer on an emotional level. It thus becomes a vital agent in communicating environmental urgency. The musician and performance—although coming from a white and Western cultural context, using an artificial ice stage and playing a music format (elegy) deeply rooted in Western culture that somehow repeats the colonial fantasies of the poles—open up space for a change in perception. Thus, Einaudi's Elegy seems to be a more sensitive approach to the topic of icy imaginaries by using the sublime as something ideologically "contaminated" (Bloom 2022, 54). Understanding the sublime as something contaminated means dealing with it in a self-critical way; as a questioning of Romantic aesthetics and their ideological heritage, rather than their mere replication. As a form of trauma work, the new, contaminated or environmental sublimity of ice approaches the world's cyrosphere as deeply shaped by an interrelated network of science and cultural perception (Bloom 2022, 75).

Today, icescapes do not fit into the narrative of regions untouched by humans. Instead, they are "irrevocably altered by remote human action and [...] will irrevocably change the course of human lives all over the globe" (Leane and McGee 2020, 1). The cyrosphere of the planet is touched by mankind, and touches our emotions when represented as disappearing phenomena in arts and media culture. In the era of the Anthropocene—a "period in which human activity [...] has become a key driving force of planetary environmental change" (Leane and McGee 2020, 3)—icy regions have moved from the margins of our imagination into the centre of our attention. Thus, it becomes crucial to examine how the emotional effects of ice are deployed, transformed and refracted in art to provide lively reflection on the complex connections between humans and nature. While Einaudi's artistic comment on environmental change and exigency is conveyed through staging, setting and visual and audible narration, it is, of course, not the only example of the power of art and aesthetics in communicating the urgency of climate action. A less overwhelming but equally effective aesthetic strategy and cultural example that conveys a message about the ongoing ice melt—thus raising awareness of its ecological processes and effects—is the story and visual history of the 'iceman', Ötzi.

Communicating Ice and Urgency in Popular Fiction

On 19 September 1991, a couple walking in the mountains on the border between Austria and Italy found a now-famous mummy, frozen in ice. The discovery—only possible due to the increasing ice melt in European mountains—quickly went viral because of his great age (more than 5000 years) and well-preserved condition. Today, scientists continue to provide new insights into Ötzi's life, appearance and death, which are then disseminated by popular media, such as magazines, television series or stamps and graphic narratives. The museum shop in Bolzano (Italy), for example, sells comics aimed at adults and children alike, in which Ötzi becomes the tragic hero of an adventurous criminal story as well as an action figure of media hype. The shop also sells a picture book for children—*Oetzi: The Iceman* in which Ötzi travels forward in time and is confronted with today's industrial society, which makes it impossible for him to lead his former life as a hunter (Bovo and Barducci 2018). In the end, Ötzi experiences many humorous events that accustom him to modern everyday life and make him aware of the similarities between today's society and his own Neolithic time period. The power of this story lies in the suspenseful narration, its sense of presence and the acute need for action, painting the picture of an individual threat that is connected to the reality of our own lives. Reflecting on the mysteries of ice, its archival function and long history, and its crucial significance for life on our planet, cultural products of this kind playfully embed complex scientific knowledge in comic book aesthetics (sequential art) and visual fiction to reach a large and diverse readership.

Climate-related popular fiction, for example in the form of comic book stories, is not only a versatile frame for interpreting our relationship with scientific themes and discourses, but also a vehicle for science communication. In other words, pop cultural media exploring science topics highlight that the public interest in science, and its perception and understanding, are embedded in a matrix of complex socio-cultural processes that give science meaning in our daily lives (Boykoff and Osnes 2019, 155). Visual science fiction reflects ideas about science and "construct[s] perceptions for both the public and scientists in a mutual shaping of science and culture" (Kirby 2008, 44); these cultural perceptions or 'cultural meanings of science' are studied in the field of science communication (among other topics). Science communicator David A Kirby clarifies: "Several studies of science popularization demonstrate that its cultural meanings, and not its knowledge, may be the most significant element contributing to public attitudes toward science" (2017, 11). Unsurprisingly, popular media and audience identification with their protagonists have been studied as a source of audience influence and as variables that can affect perceptions of (behavioural) norms and changes (Rhodes and Ellithorpe 2016, 362). Popular fiction, for example, can "reduce various forms of resistance to persuasion" by making sense of "science's nature, role, and potential" (Davies et al. 2019, 9). Provoking emotional responses in audiences can increase engagement with scientific concepts and perceptions of their usefulness or meaningfulness (Bilandzic et al. 2020), and popular fiction can raise awareness of science-related issues such as climate change (Morris et al. 2019). By creating an immersive "melding of attention, imagery and feelings" (Davies et al. 2019, 8), and fostering identification and empathy, narrative and visual fiction can influence audiences' perceptions of their own worlds (Mathies 2020; Stroud 2008). This is the 'cultural stream' of science communication in which we position *Communicating Ice through Popular Art and Aesthetics* when we reflect on science *in* and *as* culture.

About This Book

Ötzi's graphic life story and Einaudi's 'ice-breaking performance' are striking contemporary examples of the power of (popular) art, aesthetics and visual fiction to communicate environmental awareness about the loss of ice and, thus, the urgency to act—to save our planet before it is too late. Visual fiction and sublime aesthetics—two examples of artistic strategies—convey climate change in very different ways and can be both appreciated and criticised for their appropriateness and effectiveness. In this context, *Communicating Ice Through Popular Art and Aesthetics* addresses the following questions:

What artistic and aesthetic strategies are used to communicate research on ice and its crucial importance for our planet? For example, what role do heroic narratives, explorer myths and their aesthetics (the Arctic Sublime) play in our cultural imagination around ice, and how are they used to communicate the need for action in the face of climate change today?

What kind of cultural work do (popular) art and visual fiction do in the context of ice research and science communication about it?

What aesthetic achievements and social benefits are produced through the collaboration of scientists and artists exploring the urgency of ice, and in what contexts?

What role does science actually play in, and for, our cultural ideas or imaginaries of ice?

The authors of this book answer these questions by combining the individual voices of leading scientists who collaborate with artists, distinguished artists who collaborate with scientists, science communication scholars, art and cultural historians, film and media researchers and scholars from the fields of comics studies and popular entertainment studies. Removing the traditional boundaries that separate these disciplines, they produce an understanding of the cultural work—and the aesthetic and societal achievements—that emerges from the interplay between ice research, art and aesthetics and the cultural power of artistically communicated science. In so doing, *Communicating Ice Through Popular Art and Aesthetics* traces the links between broadly defined visual art forms and science (in various historical and contemporary contexts), and innovative popular visual artworks (such as cutting-edge graphic fiction and animated film) and more 'traditional' art genres (such as landscape painting/ice panoramas). It also clarifies the role of 'ice science' as a driver for (new) artistic expressions and styles, the impact of artists on new science collaborations and science communication projects, and the contribution of science communication to innovation, which is central to new understandings of cultural history. Exploring both the strategies employed in (popular) art and aesthetics to convey meaning and awareness, and how they can be made fruitful for science communication, this edited collection introduces new perspectives on how our collective environmental responsibility can be addressed and communicated across disciplines. With its focus on ice—its investigation, representation and interpretation—*Communicating Ice Through Popular Art and Aesthetics* nudges and engenders a fruitful dialogue between science, science communication, art, art history and pop cultural studies exploring visual fictions.

Research Context and Frame: Ice in Arts, Science and Popular Aesthetics

Scholarship on the political and economic interests associated with melting ice in times of climate change has flourished in recent years. While some scholars recognise the links between ice melt and cultural heritage/imagination (Dodds 2018; Wilson 2003), most publications focus on interests and conflicts in either the far North or South (Nord 2021; Evengård et al. 2015). Scholars analyse ice explorations in terms of hidden power structures in minorities and/or gender studies and discuss the role of art and artists in this context (Tennberg et al. 2022; Thisted and Gremaud 2020). However, there is a dearth of research on the possibilities and responsibilities of aesthetics and visual fiction as methods *and* results in climate change research—especially with a focus on ice melt in polar regions and other icy areas of the world, which make up the world's cryosphere.

Publications have addressed the colonial history of arctic explorations and their image policies/communication strategies in which the male and European-orientated version of the polar sublime play a major role (Kaalund 2021; McCorristine 2018). Hester Blum has examined the rich archive of polar voyages and shed light to the connection between the experiences of the Arctic and Antarctic in comparison to the increasingly extreme weather phenomena of climate change (Blum 2019). In this context, the role of ice as an active agent in environmental arts is increasingly acknowledged and examined for its historical polar exploration roots (Gould 2020). Researchers have looked at the origins, social dimensions and perceptions of ecological thinking in empirical research (Huber and Wessely 2019) and have also taken Indigenous ice knowledge and its relevance for polar films into account (McKenzie and Westerståhl Stenport 2016). Taking this as a starting point, this book offers a panoramic view and an in-depth analysis of ice as an agent not only in the context of aesthetic climate communication, but also in the context of aesthetic knowledge production.

Important publications have shed light on the relationship between the humanities and science by introducing the field of Environmental Humanities (Adamson and Davis 2017; Emmet and Nye 2017; Schmidt and Zapf 2021)—which is now divided into subcategories, including Ice Humanities. A recent publication by Klaus Dodds and Sverker Sörlin on Ice Humanities recommends a non-earth-bound perspective to foster ecological thinking, with a special emphasis on ice as metaphor for the interrelations of nature, culture and identities (2022). "As an emerging and interdisciplinary field", scholars of Ice Humanities examine "the multiple ways ice enters human and more-than-human life" (Dodds and Sörlin 2022, 1–2). Thinking about, and with, ice connects past and future; it communicates the vulnerability of our planet and the invisible forces of nature, as well as imperial endeavours and post-Anthropocene approaches. This book addresses the need for interdisciplinary research on ice by bringing together scientific contributions from the natural sciences and the humanities. We foster transdisciplinary research and shed light to the non-human agency of ice and the cultural imaginaries connected with it.

Communicating Ice through Popular Art and Aesthetics engages with publications from exhibition, performance and film studies that (implicitly) understand (Ant)Arctic ice to be a dynamic stage for visualising and communicating non-anthropocentric worldviews, and as a vital metaphor for the relationship between humankind and nature (Gough 2013; Hannah 2018; Philpott et al. 2020). Most of the authors in this volume draw from research on climate change, science communication, culture and art history, and, in so doing, tacitly agree that 'science' and 'culture' are not distinct, well-defined entities. Surprisingly little attention has been paid to this idea in the context of ice-related climate change discussions—that is, the use of ice as an aesthetic communication practice not only in visual culture, but with a special focus on art and popular culture. We address this significant gap in our understanding by considering the various shapes science takes on in an array of cultural and artistic stages, from the early times of ecological awareness to today's climate threat. Against this background, this book pioneers exploring human engagement with ice—its (re)presentation and communication—by working with ice and ice-related environmental urgency in science and science-related art, aesthetics, performance and visual fiction.

Chapter Overview

This book considers a range of perspectives, woven together to form illuminating new insights into cultural meanings of ice-related sciences and our ecological ice crisis, and their aesthetic (re)presentation and interpretation in (popular) arts. It presents a mosaic of interconnected analyses and discussions that offer a colourful and suggestive—but by no means exhaustive—picture of the interplay between scientific knowledge, aesthetic imagination and environmental communication strategies. Rather than delivering a comprehensive taxonomy of the links between *ice art* and *ice sciences*, this book zooms in on some of the spectra of their cultural halo, based on chapters that inherently 'narrate' the diversification and dynamics of ice phenomena; they inform modern sensibilities, and predate and unfold in contemporary, transhistorical, discursive, cultural continuums.

Part I of this book—"Staging Ice and Ice Stages in Science, Science Communication and Aesthetic Experience"—investigates crucial influences and interrelations between the sciences, science communication and humanities/arts, with ice as a unifying core element. The last decade has seen an upsurge in research in the arts and humanities on ice, frost, frozen ground and snow, as **Klaus Dodds** points out in his chapter on "Ice Stages and Staging Ice". In part, this reflects a determination to steer public and academic attention away from the dominant epistemologies and practices associated with the natural and life sciences. Indigenous writers have long noted that the cryosphere is an integral part of the homelands and knowledge of cold and mountainous Arctic communities. Thus, this chapter 'stages' an encounter with ice to explore and interrogate how ice and snow are enmeshed in power relations, nationalisms, heroism and heritage, settler colonialism and militarism, and how it serves as a stage for aesthetic debates, environmental activism, cultural imaginaries and Indigenous experiences. As Klaus Dodds stresses, a humanities approach to ice invites new ways of thinking and living in a world disrupted by fossil fuel capitalism and unstoppable climate change.

To minimise future harm, our societies urgently need to respond to the accelerating climate crisis, as **Craig Stevens** and **Gabby O'Connor** show in their chapter "Movies on Ice: An ArtSci Perspective on Communicating Antarctic Ice in the Climate Emergency". Despite clear evidence of a changing planet, the response across the socio-political spectrum is not fast enough. The future of the polar ice caps is a highly visible theme in this evidence-response path, both in terms of climate mechanics and in the public's awareness of 'climate science'. Much of the public imagination of ice is conveyed through media, often through images and videos that connect with the mental context of viewers. As future climate extremes are beyond our current experience, fictional films offer a connection. The authors describe an art-science collaboration around Antarctic ice-ocean field research, viewed through the lens of (experiencing) polar field camp film nights.

In his chapter "Here Be Science Show Dragons: Ice, Icons and Metaphoric Approaches to Climate Change Communication", **Graham J. Walker** explores 'science shows'—a medium that combines live science demonstrations with theatrical presentation—and their potential to communicate, entertain and transform meaning around climate change, particularly using ice, fire and other iconic representations. The author considers climate icons, such as melting ice, as entities or representations through which individuals relate to, and find meaning about, climate change. The author looks at science shows as a genre, unpacking their origins and connections to affect and motivation—including the use of ice within shows—before discussing case studies of climate change shows. He introduces considerations for climate change science shows, including empowering narratives that balance climate impacts with possible solutions; the use of icons such as ice, fire, flatulence and mangroves; and how culture and relevance shape effective iconography. The chapter questions the shortcomings and possible extensions of these shows—in particular, the potential for metaphor as a narrative device—to transform meaning and illuminate paths to positive, sustainable futures; and the risks of metaphors that fuel climate anxiety and disengage individuals.

Part II of this volume—"Ice Exploration: Heroism, Art and Imaginaries"—traces the historic roots of the heroic and the imagination of ice as an 'empty' stage in its connection to today's climate. It brings together different perspectives on the importance of ice as a geological and cultural repository and on the difficulties of understanding and communicating climate change as a processual catastrophe. The authors also examine the use of ice as a topic, performative agent, enticing metaphor and active protagonist in theatre productions and performance formats, and the ways in which they can engage different audiences with environmental messages.

In her chapter "Ethnography as Radicalised Womanhood in the Arctic Writings of Josephine Diebitsch-Peary" **Nanna Katrine Lüders Kaalund** unpacks the complex interactions between gender, race and environment in the colonial 'contact zone' of literature. In the history of polar exploration, Josephine Diebitsch-Peary is best known for accompanying her husband, Robert, in his attempts to reach the North Pole and for giving birth to their daughter while in the high Arctic. In several books such as *My Arctic Journal* (1893), *The Snow Baby* (1901) and *Children of the North* (1903) Josephine Peary described her experiences and shaped American (children's) ideas about the Arctic and the Indigenous peoples of the

Arctic. Kaalund shows that in her writings, Peary mobilised her embodied difference to the orthodox persona of the heroic male explorer as a way of garnering attention for her lecture tours and publications. By taking Peary's books as serious, significant ethnographic texts, Kaalund shows how popular literature influenced perceptions and imaginaries of extra-European peoples and environments within the context of white imperialistic expansionism.

Birgit Schneider explores the iconic status of ice within climate change communication in her chapter "Materiality of Time: Polar Ice as a Medium for Ecological Art for the Tempered Zones". Following the idea of ice as a real-time system, she offers a panorama of different artists working with ice in the critical reflection of polar explorations and polar travelling. In terms of an ecologically motivated aesthetic, these artworks allow the recipients to delve into ice as a medium. Ice allows to sensualise and imagine, but also to problematise the fragile aesthetics and artistic modes of sublimity in times of global warming.

In her chapter on "Sensing Polar Ice Bodies", **Stephanie von Spreter** investigates how contemporary artist Himali Singh Soin's long-term project *we are opposite like that* (2017–2022) engages with posthuman feminist concepts within an Arctic discourse. The author focuses on the effects of climatic changes leading to the melting of the polar ice caps and asks what the gradual disappearance of ice which had dominated the landscape and mythologies over time means. The chapter examines the disappearance of planetary history through melting polar ice, and with it the disappearance of ice as a natural archive. It discusses Astrida Neimanis's 'figuration' of bodies of water; the mythologies, ghosts and monsters left behind that remain interlocutors for our future; and the omnipresence of colonialism in the Arctic. The author asks how the relics of historical Arctic exploration still haunt us today, and how our situatedness points to our differences and distances from one another—but can also be used as a common feminist and transformative ground for creating other possible worlds.

Although Antarctica is far away, imagined versions of this place can make it accessible from afar. In their chapter "Antarctic Science on the Musical Stage" **Hanne E. F. Nielsen, Elizabeth Leane, Dana M. Bergstrom** and **Carolyn Philpott** show that Antarctica has been variously associated with heroes, extremes, purity, fragility and science. Focusing on *Antarctica—A New Musical* (2016), the authors explore how a popular stage musical work can make global science challenges accessible to a wide public and help create a connection to the ice continent for those who live far away from it. The authors analyse Antarctica's depiction on stage and the musical as a form of science communication, paying particular attention to its engagement with scientific characters and concepts. This chapter traces the evolution of the musical from a stage production to a podcast and considers how this shift speaks to both scientific and human questions around our relationship with Antarctica, and with the more-than-human world more broadly.

In "Icy Love: Performing Affect and Emotion Feeling About Climate Change", **Peta Tait** discusses the staging of ice as a particularly effective means of engaging audiences with climate change, and how ice in performance—such as *THAW* (2021) by the Australian physical theatre group Legs on the Wall—can exert a sensory appeal that encourages affects of attention. By inciting emotional feeling, performance with ice has the power to enhance viewer participation and awareness of place, while also evoking fears about the future of our planet and our environmental responsibility. By working with spectacular aesthetics, elements of visual storytelling and ice as an agent (rather than merely a stage), climate-related performance arts invite moments of encounter and stir a feeling of connection.

Part III of this edited collection—"Pop Cultural Meanings of Ice in Visual Fiction and Film"—focuses on visual science spectacles in the medium of film and comic book stories, exploring knowledge transfer and aesthetic strategies of audience activation in ecological cultural contexts.

In her chapter "Frozen Balloons: Aeronautic Heroism and Scientific Knowledge Production" **Anne Hemkendreis** examines the 2019 film *Aeronauts,* directed by Tom Harper. Here, ice and snow function as poetic storytelling tools, strategies of emotionalisation and, above all, key elements of meteorological research. The biopic of pioneering meteorologist James Glashier includes a modern fictional heroine, aeronaut Amelia Wren, who highlights the ignorance of women's roles in ice research and history. In doing so, the film draws attention to the gaps in understanding, researching and communicating environmental conditions. Contrasting science and show business, the film draws attention to humans' entanglement with nature's visible and invisible forces. This chapter explores the role of ice as an element within, and evidence of, scientific knowledge production. It shows how ice is used in popular culture to challenge our understanding of science, to communicate environmental knowledge and to question our relationship with nature in times of climate change.

Johannes Riquet concentrates on two other films in his chapter "Hard Ice, Soft Snow? Transnationalism, Spectatorship and the Arctic Sublime in *Chasing Ice* (2012) and *Silent Snow* (2011)". His comparative analysis of two environmental films uncovers that *Chasing Ice*, a climate-change documentary, reimagines the Arctic sublime by exploring climate change as a visual spectacle recorded by a photographer-turned-heroic Arctic explorer. In contrast, *Silent Snow* links art and activism in a transnational aesthetic based on relational experience, rooted in the Inuit understanding of snow and ice. Riquet argues that both films are about communicating and visualising environmental issues, dealing with snow and ice both materially and metaphorically. However, they do so in completely different ways, thus making the ignorance within the most dominant and persistent (heroic) imaginations of the poles visible. This chapter shows that Indigenous knowledge about ice—often considered inferior to Western scientific knowledge production—is, in fact, indispensable and extremely profound for a changed and healthier human relationship with the planet.

Frédéric Bouchard and **Ylva Sjöberg** take a different approach in their chapter "*Frozen Ground Cartoons*—Revealing the Invisible Ice", which focuses on permafrost—the frozen ground that occupies more than 20 million square kilometres of the Earth's high-latitude and high-altitude landscapes. The authors argue that permafrost is not only a key component of our global climate, but the most overlooked invisible component of the cryosphere. However, just like the often-media-depicted ice sheets in Greenland and Antarctica, or mountain glaciers around the world, permafrost regions are extremely sensitive to climate change and human activities, which is why permafrost research is of great scientific and societal importance. The "Frozen-Ground Cartoons" project was born from the need to make permafrost science accessible and fun for the general public, especially for school kids, students and teachers. A booklet of comic strips was the original outcome; however, the project quickly evolved into a series of 'by-products', including translations into several languages, augmented reality materials (maps, photos, videos, 3D drawings), a board game, etc.

While scientific data is often seen as the dominant expression of research, including research on ice, public understanding and engagement are embedded in a matrix of complex (cultural) processes that give ice meaning in our daily lives. Popular culture—in the form of comic books and *bande dessinées*—is a versatile frame for interpreting our relationship with scientific and environmental discourses as **Anna-Sophie Jürgens, Stefan**

Buchenberger, Laurence Grove and **Matteo Farinella** show in their chapter "On the Visual Narratives of Ice in Popular Culture: Comics on Ice, Icy Villains and Ice Science". Visual narratives play an important role in shaping our cultural ideas of ice, and this chapter explores their many facets in 'proto-comics' of early-modern emblem books, the first modern comics, early twentieth-century sequential art fantasies and contemporary *bande dessinées*. Bringing together perspectives from science communication, comics studies and popular entertainment studies, the authors focus on contemporary examples of 'comics on ice' and comic book ice science villains to highlight how the intrinsically hybrid and changing nature of sequential art—and its ability to visually express non-visual emotions—can help us imagine the unimaginable (ecological futures) and define what might be called the 'visual narratives of ice'.

The melting of the polar ice shields that protect our planet and stabilise vast parts of the Earth's biosphere has been widely recognised across the planet. This ice crisis—as **Ben Nickl** makes clear in his chapter "Melt for Me: Communicating Ice Empathy Through the Plasticity of Disney"—has created a cryocritical awareness of the retreat of ice bergs, ice shelves and glaciers as a direct result of human action. By building on concepts of animation and the plasticity of feeling, the chapter presents a case study of ice in Walt Disney's 2013 feature film *Frozen* as a popular plasmo-affective feel-view language. The author argues that with ice animates such as Olaf the snowman, we are emotionally reshaped, bent and stretched, as we move closer to imagining and intuiting the enormous importance of the ice in our world—and the need for humans to develop a new emotional relationality with it. As Nickl shows, *Frozen* evokes feelings of connections with ice and opens up a room for a changed relationship between humans and the planet.

To bring *Communicating Ice Through Popular Art and Aesthetics* full circle, the last chapter—"On the Aesthetic Facets of Ice Urgency: Some Final Reflections" by **Anne Hemkendreis, Anna-Sophie Jürgens** and **Karina Judd**—recaps the core themes from this volume and critically reflects on the definition of key concepts, such as 'communication', 'ice art' and 'stages'. The final chapter examines the research gaps in our cultural understanding of ice through (popular) arts and aesthetics, and the interdisciplinary breadth added to the discussion by the authors of this collection. In doing so, we reflect on what the voices from various disciplines in this book define as 'communication with ice' in science-related arts popular aesthetics. Finally, this book considers the reciprocal influence and mutually imparted meaning of ice and arts, showing how scholars and artists can inspire the use of ice as an active agent within contemporary and future artworks dealing with climate change.

Bibliography

Adamson, Joni, and Michael Davis. 2017. *Humanities for the Environment. Integrating Knowledge, Forging New Constellations of Practice*. London: Routledge.

Bilandzic, Helena, Susanne Kinnebrock, and Magdalena Klingler. 2020. The Emotional Effects of Science Narratives: A Theoretical Framework. *Media and Communication* 8 (1): 151–163. https://doi.org/10.17645/mac.v8i1.2602.

Bloom, Lisa. 1993. *Gender on Ice. American Ideologies of Polar Expeditions*. Minneapolis: University of Minnesota Press.

———. 2022. *Climate Change and the New Polar Aesthetics: Artists reimagine the Arctic and Antarctic*. Durham: Duke University Press.

Blum, Hester. 2019. *The News at the Ends of the Earth. The Print Culture of Polar Exploration*. Durham: Duke University Press.

Boetzkes, Amanda. 2010. *The Ethics of Earth Art*. Minneapolis: University of Minnesota Press.

———. 2020. Climate Aesthetics in the Ablation Zone. Afterimage. *The Journal of Medial Arts and Cultural Criticism* 47 (2): 35–39. https://doi.org/10.1525/aft.2020.472007.

Bovo, Eleonora Suri, and Armin Barducci. 2018. *Ötzi. The Iceman*. Bozen: Athesia.

Boykoff, Maxwell, and Beth Osnes. 2019. A Laughing Matter? Confronting Climate Change Through Humor. *Political Geography* 68: 154–163. https://doi.org/10.1016/j.polgeo.2018.09.006.

Brady, Emily. 2013. *The Sublime in Modern Philosophy: Aesthetics, Ethics and Nature*. Cambridge: Cambridge University Press.

Davies, Sarah Rachael, Megan Halpern, Maja Horst, David Kirby, and Bruce Lewenstein. 2019. Science Stories as Culture: Experience, Identity, Narrative and Emotion in Public Communication of Science. *Journal of Science Communication* 18 (5): A01. https://doi.org/10.22323/2.18050201.

Dodds, Klaus. 2018. *Ice. Nature and Culture*. London: Reaktion Books.

Dodds, Klaus, and Sverker Sörlin. 2022. *Ice Humanities. Living, Thinking and Working in a Melting World*. Manchester: Manchester University Press.

Einaudi, Ludovico. 2016. *Elegy for the Arctic*. https://www.youtube.com/watch?v=2DLnhdnSUVs. Accessed 01 Dec 2022.

Emmet, Robert S., and David E. Nye. 2017. *The Environmental Humanities. A Critical Introduction*. Cambridge, MA: MIT Press.

Evengård, Brigitta, Joan Nymand Larsen, and Øivind Paasche. 2015. *The New Arctic*. Berlin: Springer.

Glasberg, Elena. 2011. "Living Ice": Rediscovery of the Poles in an Era of Climate Crisis. *Women's Studies Quarterly* 39 (3/4): 221–246. https://doi.org/10.1353/wsq.2011.0072.

Gough, Richard. 2013. *On Ice*. Performance Research: Special Issue 18 (6), Routledge.

Gould, Polly. 2020. *Antarctica. Art and Archive*. London/New York: Bloomsbury.

Grawe, Johannes. 2001. *Caspar David Friedrich und die Theorie des Erhabenen. Friedrichs "Eismeer" als Antwort auf einen zentralen Begriff der zeitgenössischen Ästhetik*. Kromsdorf: VDG.

Hannah, Dehlia. 2018. *A Year Without a Winter*. Bethesda: Columbia Books.

Horn, Eva, and Hannes Bergthaller. 2019. *The Anthropocene: Key Issues for the Humanities*. London: Routledge.

Huber, Florian, and Christina Wessely. 2019. *Milieu. Umgebungen des Lebendigen in der Moderne*. Paderborn: Fink.

Kaalund, Nanna Katrine Lüders. 2021. *Explorations in the Icy North. How Travel Narratives shaped Arctic Science in the Nineteenth Century*. Pittsburgh: University of Pittsburgh Press.

Kirby, David. 2008. Cinematic science. In *Handbook of Public Communication of Science and Technology*, ed. Massimiano Bucci and Brian Trench, 41–56. London: Routledge.

———. 2017. The Changing Popular Images of Science. In *The Oxford Handbook of the Science of Science Communication*, ed. Kathleen H. Jamieson, Dan M. Kahan, and Dietram A. Scheufele, 291–300. Oxford Handbooks Online.

Leane, Elizabeth, and Jeffrey McGee. 2020. Anthropocene Antarctica: Approaches, Issues and Debates. In *Anthropocene Antarctica: Perspectives from the Humanities, Law and Social Sciences*, ed. Elizabeth Leane and Jeffrey McGee, 1–14. London: Routledge.

Mathies, Susanne. 2020. The Simulated Self—Fiction Reading and Narrative Identity: 'How Can I Have a Complete Identity Without a Mirror?' (William Golding, Pincher Martin). *Philosophia* 48 (1): 325–345. https://doi.org/10.1007/s11406-019-00079-3.

McCorristine, Shane. 2018. *The Spectral Arctic. A History of Dreams & Ghosts in Polar Exploration*. London: UCL Press.

McKenzie, Scott, and Anna Westerståhl Stenport. 2016. *Films on ice. Cinemas on the Arctic*. Edinburgh: Edinburgh University Press.

Morgan, Benjamin. 2016. After the Arctic sublime. *New Literary History* 47 (1): 1–26.

Morris, Brandi S., Polymeros Chrysochou, Jacob Dalgaard Christensen, Jacob L. Orquin, Jorge Barraza, Paul J. Zak, and Panagiotis Mitkidis. 2019. Stories vs. Facts: Triggering Emotion and Action-Taking on Climate Change. *Climatic Change* 154 (1–2): 19–36. https://doi.org/10.1007/s10584-019-02425-6.

Neimanis, Astrida. 2019. *Bodies of Water. Posthuman Feminist Phenomenology.* London/New York: Bloomsbury.

Nord, Douglas C. 2021. *Nordic perspectives on the responsible development of the Arctic: Pathways to action.* Berlin: Springer.

Philpott, Carolyn, Elizabeth Leane, and Matt Delbridge. 2020. *Performing Ice.* London/New York: Palgrave.

Plumwood, Val. 1993. *Feminism and the Mastery of Nature.* London: Routledge.

Randerson, Janine. 2018. *Weather as a Medium. Toward a Meteorological Art.* Cambridge: MIT Press.

Rautmann, Peter. 1991. *Caspar David Friedrich. Das Eismeer. Durch Tod zu neuem Leben.* Frankfurt am Main: Fischer.

Rhodes, Nancy, and Morgan E. Ellithorpe. 2016. Laughing at Risk: Sitcom Laugh Tracks Communicate Norms for Behavior. *Media Psychology* 19 (3): 359–380. https://doi.org/10.1080/15213269.2015.1090908.

Schmidt, Matthias, and Hubert Zapf. 2021. *Environmental Humanities. Beiträge zur geistes- und sozialwissenschaftlichen Umweltforschung.* Göttingen: V&R Unipress.

Schneider, Birgit. 2021. Sublime Aesthetics in the Era of Climate Crisis? A Critique. In *The Routledge Companion to Contemporary Art, Visual Culture and Climate Change*, ed. T.J. Demos, Emily Eliza Scott, and Subhankar Barnerjee, 263–273. London: Routledge.

Scholz, Tobias. 2021. *Distanziertes Mitleid: Mediale Bilder, Emotionen und Solidarität angesichts von Katastrophen.* Campus.

Stroud, Scott R. 2008. Simulation, Subjective Knowledge, and the Cognitive Value of Literary Narrative. *The Journal of Aesthetic Education* 42 (3): 19–41. https://doi.org/10.1353/jae.0.0016.

Tennberg, Monica, Else Grete Broderstad, and Hans-Kristian Hernes. 2022. *Indigenous Peoples, Natural Resources and Governance. Agencies and Interactions.* London: Routledge.

Thisted, Kirsten, and Ann-Sophie N. Gremaud. 2020. *Denmark and the New North Atlantic. Narratives and Memories in a Former Empire.* Aarhus: Aarhus Universitetsforlag.

Weber, Gary R. 2018. The Popular Culture of Disaster: Exploring a New Dimension of Disaster Research. In *Handbook of Disaster Research*, ed. Enrico L. Quarantelli and Russel R. Dynes, 430–440. New York: Springer.

Wilson, Eric G. 2003. *The Spiritual History of Ice. Romanticism, Science, and the Imagination.* London/New York: Palgrave.

Ice Stages and Staging Ice

Klaus Dodds

Introduction

Global media attention is increasingly preoccupied with ice's diminishing state and vanishing volume. Numbers and predictions abound and much of it revolves around implications for sea level change. The world's ice stage is under siege from excessive heat and moisture, made worse by a warming and expanding ocean. The emergence of a self-conscious Ice Humanities, with parallels to Blue and wider Environmental Humanities, is a timely reminder that human and non-human history does not stop at the edge of land (Bloom 2022; Cosgrove and Della Dora 2008; Dodds 2021; Dodds and Sörlin 2022; Leane 2020; Ruiz et al. 2020). Ice—whether in glacial bodies or ice sheets, underwater, or on top of land, lakes and seas—has never been divorced from wider cultural, ecological and material interactions and spaces. In mainstream media, it is commonplace to read about the precarious state of ice and its association with accelerating climate change and environmental state-change. Ice and snow are

K. Dodds (✉)
Royal Holloway, University of London, Egham, UK
e-mail: K.Dodds@rhul.ac.uk

A. Hemkendreis, A.-S. Jürgens (eds.), *Communicating Ice through Popular Art and Aesthetics*, Palgrave Studies in Media and Environmental Communication,
https://doi.org/10.1007/978-3-031-39787-5_2

storified and amplified through emotive plots and narrative denouement. A BBC Radio 4 series entitled *The Dying of the Ice* captured a particular elemental and even solastalgic mood, framing ice as vulnerable and precarious (BBC 2019). The Thwaites glacier is routinely framed as the 'Doomsday Glacier' because of its capacity to surge and splurge.

As the world gathered in Glasgow to discuss the fate of the Paris Agreement in November 2021, news broke that a series of glaciers in Antarctica were going to be named after cities and islands that have hosted major climate change summits. The Antarctic map now features a 'Glasgow Glacier' in West Antarctica close to the Thwaites Glacier. According to the European Space Agency, the nine so-called climate glaciers were identified for naming based on "using data from satellites, was found recently to have lost more than 300 gigatonnes of ice over the last 25 years" (ESA 2021). This is not the first time glacial ice has played a part in climate negotiations and their mediation. The Icelandic-Greenlandic artist, Olafur Eliasson, curated an *Ice Watch:* twelve blocks of Greenlandic ice were deposited outside the conference centre in Copenhagen during the Conference of Parties (COP) 15 negotiations. Participants and spectators were invited to watch the ice slowly melt into the urban infrastructure of the Danish capital city (for an elegant interrogation, see Hosseini 2019). This was, in essence, an affective and embodied encounter, as some spectators wished to touch, taste and smell the melting ice. The artist noted at the time that this sensual attraction was integral to knowledge communication; audiences learnt through a direct and tangible experience that the world's ice was a melting stage.

As a suite of anthropologists and indigenous scholars have noted, ice is interdependent and entangled with human and more-than-human communities (e.g. Cruikshank 2005; Tall Bear 2017). For millennia, communities in the Arctic, ice-covered mountainous regions and other cold places have lived, moved and worked with ice (Lincoln et al. 2020; Watt-Cloutier 2015). But the capacity to live in ice-covered regions of the world continues to be sorely tested by a suite of circumstances and forces that originate in faraway places. Black carbon, for example, is a by-product of wildfires sweeping the temperate and boreal parts of the world, and is a major contributor of ice melt. Deposited on the Greenlandic interior ice and elsewhere in the Arctic region, black carbon effectively acts as a source of unwelcome warmth to the underlying ice. The Arctic's albedo—its capacity to reflect solar heat back into the atmosphere—is diminishing as it becomes brown, green and even black due to wildfire debris landing on its glacier and ice fields.

In geopolitical discussions of ice and snow, the focus has often been concentrated on how they act as barriers to human mobility and operability—challenges to the functioning of geopolitical power and the territorial reach of the State (Dodds 2018). Ice bodies have been an object of conflict, as is the case of India and Pakistan's struggles over high-altitude environments. But ice can also become a national security issue; melting ice is an essential source of water supply for millions in South America, Central and South Asia and many other parts of the world, which means border wars are more likely as states struggle with the paradoxical qualities—too much at times and too little at other times—of that meltwater (Dodds and Smith 2022). It is not uncommon to read of glaciers being framed as 'water towers' with associated concerns regarding access and control (Elmore et al. 2020).

Nowadays, ice is increasingly recognised as an essential resource (when melted or not) and integral to human and more-than-human life. Physical and life scientists have drawn attention not only to the role that ice plays in the regulation of global climate but also in the complex ecosystems that ice supports, and that within ice there is much evidence of micro-organic life. The underside of sea ice, glacial crevasses and river ice are fertile spaces for benefiting and prospering from the materiality and spatiality of ice thickness, cover and disposition. More unsettlingly, ice melt is revealing disturbing implications as the release of poisonous substances—such as heavy metals and mercury—make their way through, and into, surrounding landscapes and seascapes. The release of diseases such as anthrax, due to permafrost thawing, reveals how the elemental intersection of ice, land and water brings with it even more unwelcome legacies. More poignantly, the Svalbard Global Seed Vault (opened in 2008 and designed to preserve and protect the world's seed varieties from future calamity) suffered flooding in 2017 when the underlying permafrost thawed. Three years later, the Vault re-opened after substantial investment in waterproofing by the host, the Norwegian government, but the reality that the ground is not reliably frozen poses questions about whether the vault is fail-safe in the face of ongoing warming.

Glaciers, as recent staged funerals in Iceland and Switzerland remind us, have become a Rosetta stone for the Anthropocene (Jackson 2019; Schmidt 2021). They simultaneously act as a metaphor for accelerating melting and disappearance due to a warming world, an observable and traceable landscape for climate change science, a potent form of environmental media and a framework from which activists, indigenous communities and governments have demanded greater legal and political protection. Working with ice, if you will, starts with recognising that changes involving ice may appear on any scale, from the individual to the continental to the global. If ice preserves past life, it is also complicit with our artificially chilled lifestyles in the here and now. If ice is recognised as a geological force, then let us also acknowledge the extraordinary ecological service it performs for the planet by moderating the power of the sun. If ice is a 'prop' for illustrating the urgency of the climate emergency, then it is also a recorder of our pre-industrial pasts (Antonello and Carey 2017). The fate of ice is intimately linked to ongoing climate change and the spectre of rapid, catastrophic and irreversible ice loss is not beyond the realm of short-term possibility. Melting, disappearance and flooding take their toll on communities and countries in uneven ways—the disastrous flooding affecting Pakistan in the summer of 2022 is a painful reminder of what is at stake as vast swathes of the country were inundated.

In this chapter, ice stages and staging ice act as a starting point to think and act with ice. In the first section, I explicitly consider the idea of the stage, including how we might think of ice as stage, staging and stage-nation. Thereafter, my attention turns to ice as a 'crisis' concept which has problematised the staging of ice and highlighted, for example, the potentiality of ice to criticise and challenge the staged 'settler cryosphere' in the Arctic. Finally, I conclude with the challenge of staging mass melting, and how to engage with a world where ice continues to be re-packaged, re-purposed and re-staged as part of an ongoing reckoning with, and resistance to, carbon-intensive activities and practices that do not align with sustainable living. Overall, I outline the history of snow and ice through a humanities lens, selecting chronological highlights that illustrate our contested relationship to this ephemeral material.

Ice Stage and Ice Stages

For centuries, Euro-Western exploration of the Arctic and Antarctic was fixated with explorers going ever further onwards and inwards (Dodds and Smith 2022). In the Arctic, the search for the fabled Northwest Passage proved to be a powerful stimulus for a medley of expeditions and sponsors (Craciun 2016). The 'magnetic' pull of the North was potent and there was no shortage of commentary about a mythical 'open Polar sea' that might greet intrepid explorers brave enough to penetrate the initial areas of sea ice. Disaster befell many of those who sought fame and fortune through and beyond the ice. As a host of authors, including Lisa Bloom (1993), identifies, the ice itself was integral to the staging of these colonial and imperial adventures and conquests (also Bloom 2022). The race to the Antarctic Pole in 1911–1912 is one of the most interrogated moments in that engagement. Images of Scott's and Amundsen's respective parties captured a vision of white men struggling to overcome the alien environment of the coldest place in the world. Surrounded by national flags, the connection between ice, men and nation was deliberately cultivated by explorers, sponsors and domestic publics. A failure to return was no barrier to commemoration and memorialisation, which matters even more on a continent—devoid of an indigenous population—that does not retain permanent traces of occupation and belonging (Bloom et al. 2008). The Captain Scott party and the Terra Nova expedition were immortalised through memorial, public appeals, films and the establishment of the Scott Polar Research Institute at Cambridge University (Jones 2004).

The staging of ice and nationalism continued throughout the twentieth century as the Antarctic became ever more coveted by claimant states and territorial rivals (Fig. 1). Men were dispatched by ship, plane and motorised vehicle to evermore regions of the Polar continent and encouraged to deposit flags, establish bases, map territory, evaluate resources and undertake scientific research around the thickness of the ice cap, the size and geological structure of the continent and the biodiversity of landscapes and seascapes. This activity was dutifully recorded, sketched, filmed and photographed. Communicating exploration, science and geopolitical activity (such as flag-planting) was essential as governments and publics were unlikely to be first-hand visitors. Territorial, resource and scientific rivals also needed to be informed about these activities and staging one's activities on the ice was integral to what one might think of as a settler colonisation project (Mancilla 2021). Integrating an uninhabited continent into a national territorial-cartographic imagination required intellectual and material labour, some of which included extracting ice, rock and indigenous flora and fauna, and transporting them back to centres of calculation and assessment in distant capitals (Roberts 2011).

For the first half of the twentieth century, the ice stage of Antarctica was one dominated by exploration and colonisation. By the 1940s, the Antarctic was the subject of seven territorial claims by Argentina, Australia, Chile, France, New Zealand, Norway and the UK. The two emerging

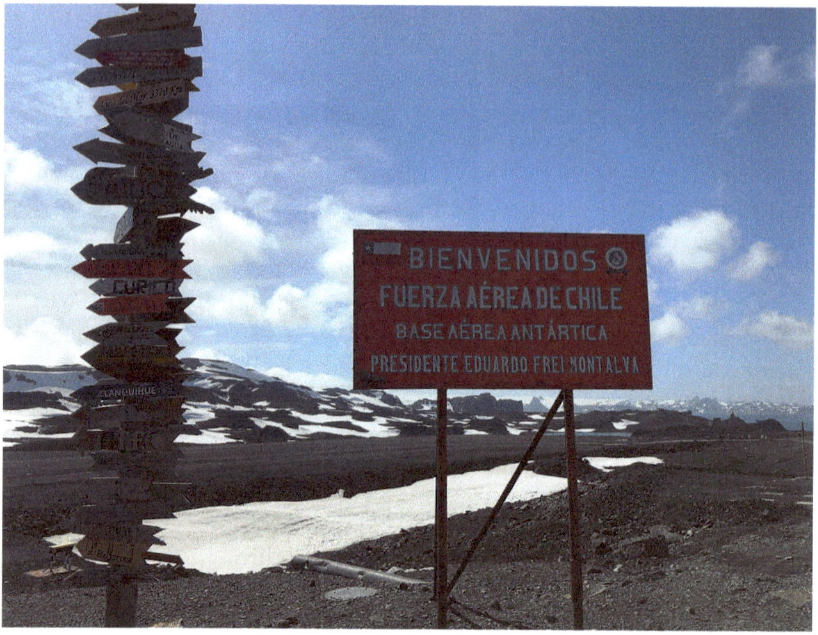

Fig. 1 Base Presidente Eduardo Frei Montalva is a Chilean base located on the Fildes Peninsula in King George Island, South Shetland Islands, Antarctica. It is located on an ice-free region and is part of the self-described Chilean Antarctic Territory, 2022 © Klaus Dodds

superpowers, the US and the USSR, reserved the right to make their own claims in the future. The seven claimant states and two superpowers did not mutually and formally recognise one another's claims; Antarctica was a contested ice stage. Within four decades, there was a shift from a lonely group of men planting their national flags at the South Pole to a situation where new maps and lines were being established on a continent that was, itself, barely mapped and charted. The ice was proving no barrier to epistemic blank spots and geopolitical ambition. While the emergence and introduction of the 1959 Antarctic Treaty has often been noted as tempering such ambitions, the framework was designed to protect the interests of the signatories, and curtailed rapid introduction of new members.

During the Cold War years, ice and snow were subject to considerable investment and scrutiny. Systematising knowledge of the ice stage was connected to power projection, infrastructure development and exploitation of frozen earth and ocean. The Arctic and Antarctic were probed, drilled and interrogated for evidence of past climate, contemporary weather patterns and future strategic value. In Greenland and Antarctica, ice cores were extracted and archived elsewhere. Scientific stations were extended and expanded across the Polar continent, and new material connections were forged between ice, ocean and atmosphere. The 1957–1958 International Geophysical Year has often been credited with this elemental transformation—publicly communicating the links between ice and the oceans, air and even outer space. But the 'Cold War' was aptly named. Ice, snow, permafrost and ice cores were integral to the military-industrial-academic complex on both geopolitical sides of the Arctic Ocean. To the Americans, the northern latitudes were hugely important for hemispheric defence purposes. Soviet bombers might fly over the ice cap and enter North American air space, as the Bering Strait divided the USSR and Alaska by less than a hundred kilometres.

Understanding Polar meteorology was one element to this northern recasting, but so was ensuring that the United States had suitable infrastructure in place to cope with the ice, cold and darkness of the High North (Farish 2010). Later there were fears that the Soviet Union might send their nuclear-powered submarines under Arctic ice and emerge in Hudson Bay to launch ballistic missiles towards the United States. Greenland was a notable 'cold spot' for military and scientific investment (Doel et al. 2016). Danish, Swiss and US glaciologists used the massive engineering infrastructures in the Camp Century site in northwest Greenland to produce the tallest ice core to date, almost 1400 metres. While scientists were extracting ice cores from the inland interior and sending the cores to the US Army's Snow, Ice and Permafrost Research Establishment (later the Cold Regions Research and Engineering Laboratory, CRREL), the US military was hatching plans to hide missiles under the ice at Camp Century via Project Ice Worm. Knowledge and understanding of ice mechanics and physics was integral to scientific and military agendas, and as far as funders (such as the Office of Naval Research) were concerned, cold weather engineering and ice studies in general were well aligned (Herzberg et al. 2018). Both Projects Ice Worm and Camp Century were later abandoned because the subglacial base camp stage was being crushed by glacial force and plasticity.

While scientists were learning how to extract ever longer ice cores from Greenland in the 1950s and 1960s, satellite technology extracted data from a great height, rather than depth. The 'satellite era', dating from the 1970s onwards, is a game-changer for how sea ice can be charted and even understood in terms of distribution, area and behaviour. However, aerial survey is not replaced by satellites, as the recent NASA's Operation Ice Bridge missions over Greenland and Antarctica demonstrate. However, satellites do allow for increasing precision in capturing sea ice minima and maximums. Underwater data collection is not redundant either, because satellite coverage struggles to collect information about ice volume and thickness. Submarines, and later drones, have been essential for that volumetric work. Thus, infrastructures and resources are integral to knowledge production of ice, raising important questions about who gets to generate that remote sensing; this, in turn, provides opportunities for other Polar powers, such as China and the European Union, to contribute to public and scientific debates about the state of ice.

The growth and development of remote sensing lends itself well to not only the rise of systems modelling but also the public visualisation of ice. In the 1980s and 1990s, the Arctic and Antarctic were popularised and visualised through charts, maps and imagery. While the Brazilian-hosted Earth Summit of 1992 (which was designed to enhance the profile of environmental issues in world politics) had the Amazon rainforest as its immediate material and visual backdrop, it was the trio of jungles, poles and mountains that were increasingly recognised as proxies for global environmental change, with the ice core acting as a touchstone of the Anthropocene (Alley 2002). Looking backwards, the 1980s was a pivotal decade of change for the cryosphere. Until the 1980s, for example, Greenland rarely experienced any summer season melt at all. What started at the lower latitudes became evident in the highest areas of the interior ice. Strikingly, environmental campaigning was becoming more vociferous around protecting the Polar regions from mining and waste management, rather than loss or disappearance per se. Notably, however, the first Intergovernmental Panel on Climate Change (IPCC) report (1990) remarked upon ice loss and the likely impact on ecosystems and communities in light of 'global warming'. The report clarified the prevailing scientific understanding of the Earth's ice sheets at the time: they were stable for the foreseeable future. Because of this, more pressing matters—like the rapid deforestation of the Amazon—gained more popular and political attention than the cryosphere.

In the last two decades, ice has been widely recognised as being in crisis. In 2002, the splintering and collapse of the Larsen B ice shelf turned 'slow thinking' upside down, providing powerfully charismatic imagery of rapid collapse even as experts claimed the disintegration was not causally linked to anthropogenic climate change (O'Reilly 2017). No longer thought of as a protective envelope, impenetrable barrier or a source of frustration to human design and mobility, the focus on ice today is resolutely on shrinkage, disappearance, retreat and thawing. A new generation of satellites and drones are monitoring the state of the world's ice sheets and glaciers and newer actors, such as China, are now substantive 'cryosphere powers' with their own ice-watching powers. Ice is, and will remain, hyper-politicised and culturalised—there is no shortage of commentaries on the unsettled legal status of ice and speculative plans to tow icebergs from Polar regions to water-stressed cities and environments.

Beyond that extractive Anthropogenic logic, ice is also caught up in a wholesale production of narratives, impacting communities living with, and affected by, its disappearance. Interlinking the local and the global was addressed by the idea of 'Planetary Boundaries', which was first published in *Nature* in 2009 with many subsequent follow-up studies (Rockström 2009). It argues that planetary stability can be subsumed under nine critical dimensions—such as rate of species loss, CO_2 levels in the atmosphere, ocean salinity and six others—with quantified boundaries. Each of the boundaries is defined against a background of previous change, and conclusions are drawn from their past 'performance'. Their timescales of change differ wildly, including in pre-Anthropocene, or even pre-human, times. The now-emerging Anthropocene awareness builds significantly on environmental patterns identified and investigated in the past. Such organisational devices are abundant in Anthropocene research where they demonstrate—often in sharp, sometimes baffling, detail—how environmental times are predicated on human agency.

The 2007–2008 International Polar Year, following on from the 1957–1958 International Geophysical Year, emphasised the importance of ice knowledge to both scientists and indigenous communities. Community participation, especially in the Arctic, was integral to the programme, as local communities expressed concerns about the Arctic and non-Arctic states' desire to extend ever greater control over open land and water, and over-determination of what activities should take place on land and underwater. There is a residual danger that a new sort of 'settler cryosphere' takes hold, where the powerful and privileged materially and epistemologically dispossess those with traditional indigenous knowledge and experience. One can see this, for example, in the way celebrity campaigners—from

Bridget Bardot to Emma Thompson to Paul McCartney—have travelled to the Canadian sub-Arctic and Svalbard to draw attention to the practice of seal hunting and/or the apparent absence of ice. The 'loss' of ice acts as a proxy to justify eco-anxiety and accompanying demands for action at a time when indigenous peoples are demanding and securing their rights to consultation, land, resource development and self-determination. The ice stage is continually contested and politically charged.

The Ice Stage in Crisis

The Ice Humanities 'turn' is a means of staging an encounter with the temporality, materiality and spatiality of ice. This includes a fundamental recognition of indigenous and settler communities and their experience and knowledge of ice and snow, whether that be in the Arctic or other cold and mountainous areas. It also generates global conversations and encounters that do not solely associate ice and snow with emotional allure but settler communities in cold and high places.

This is particularly relevant when recognising the extraordinary importance of indigenous experience, memory and knowledge of ice, which runs into millennia. Yet, ice-monitoring techniques—such as repeat photography of glaciers—paid little to no attention to local communities and their knowledges and experiences of that remote-sensed ice (Carey 2010). Indigenous communities have been working with the conceptual and embodied terrain long before glaciologists or ice humanities scholarship encountered it (Hastrup and Olwig 2012). As such, conventional ice metrics cannot do justice to lively and animated encounters with ice, as Julie Cruikshank revealed in her work with Athapaskan and Tlingit oral traditions in the Pacific Northwest. Glaciers are sentient. Ice is animate. Glaciers are responsive to encounters with humans. As Cruikshank noted, "the glaciers listen, pay attention, and respond to human behaviour" (2001, 378; 2005).

Indigenous and local communities living with ice have long favoured a view of it as capricious and capable of being 'calmed' with respectful action. In northwest India, for example, local communities recognise that they must care for glaciers and take care of them while nurturing their own spiritual interactions with ice and each other (Gagné 2019). As humanities and social science scholars have emphasised, making sense of climate change is too important to be left to the environmental and physical sciences alone. While these disciplines can detect the melting, the species loss and the micro-plastics (in places where we thought they should not be, such as the Central Arctic Ocean), it is indigenous communities who are bearing the existential brunt of ice loss, permafrost thawing and what notable indigenous Canadian activist Shelia Watt-Cloutier (2015) described as the "right to be cold".

Living with ice does not just involve human communities. From microbial life in and around glaciers, to animals and plants that depend on ice and snow, taking ice seriously necessitates reflecting upon our (in the broadest sense) co-dependencies and kinship. Without ice our planet would be radically different; its absence is as significant as its presence. Ice is integral to global climate systems and the global circulation of water and air. The surface albedo of earth is critically dependent on ice; ice reflects, rather than absorbs, solar radiation. Around the world, hundreds of millions of people depend upon the spring melt of glaciers for essential water supplies, reaching from mountain villages to farming families on the plains and deltas of Indus, Ganges and Brahmaputra. China's geopolitical interest in Tibetan and other 'Third Pole' glaciers is growing (Zhang et al. 2019). The transition of ice to water is integral to well-being, but also hazardous to human communities if glacial bursts and flash flooding follows. How we tell stories about ice reveals our relationship to ice, but also our capacity and willingness to listen and learn from other story-telling communities. And depending on who the 'we' refers to, it opens-up a trans-disciplinary, cross-cultural, multi-spatial and temporal conversation involving different oral, written and artistic forms of communication and expression. This has proven integral to indigenous-scientific-anthropological projects involving ice and the more-than-scientific communication of climate and community change (Gearheard et al. 2013). These sorts of collaborations serve as a powerful reminder that we cannot understand and make sense of ice and snow in cultural, historical, political and epistemological isolation.

Ice is neither ahistorical nor—as we are discovering, to our detriment—is it impervious to intensifying anthropogenic climate change. Ice needs to be understood as a 'crisis concept', rather than a 'state of crisis', which simply invites us to witness its disappearance and fragility. There is a need to shift the focus away from a simplistic 'disappearing optic' around ice that is easy to see—such as sea ice and glacial ice melt—to harder-to-visualise underground ice—such as permafrost. This optic privileges Euro-American perspectives on ice science and ice metrics. As a 'crisis concept' par excellence, a focus on ice—rather than, say, 'environment'—heralds a shift in the scale and intensity of human-earth historiography (Dodds and Smith 2022).

A focus on elemental substances such as ice, rock and water signals two major fractures. Firstly, there has been a move to 'decolonise' the earth and physical sciences that underpin dominant Euro-Western epistemological frameworks, heroic explorer-scientist figures, European timescales and framings such as 'ice age'. Earth sciences (e.g. geology) and specialist fields (e.g. glaciology) have been critiqued for their complicities with settler colonialism and scientific racism. One of the leading figures of modern glaciology, the Swiss scientist Louis Agassiz, combined an interest of earthly ice ages alongside an advocacy of racist and racial polygenism (which posits that the human species has multiple origins) and anti-Darwinian creationism. The European Geophysical Union renamed its Louis Agassiz Award (named in 2005) to the Julia and Johannes Weertman Medal in 2019. Marvelling at ice coincided with ideologies that advocated Western supremacy, and the invention of energy concepts and technologies that were (and are) complicit with ice's accelerated disappearance. For example, at about the same time Agassiz was writing his pioneering book about glaciers (*Études Sur Les Glaciers*, 1840), the use of coal-fired engines ushered in a new steam era. Decolonisation is ongoing and the forces that extract resource value from land, ice and sea are still very much part of a terra-altering system otherwise known as capitalism, geopolitics and settler colonialism (Grove 2019). Secondly, ice has now been recognised as a 'matter of concern' and even a 'concerning matter' rather than an object of curiosity or inconvenience. It is an 'environmental object' that is now a matter of, and for, environmental and legal governance, in the form of legislation (to protect it) or treatment (as if it is a patient—which it is, in a sense, as it slides ever deeper into the crisis).

To become a 'crisis concept' (as argued in Dodds and Sörlin 2022), ice first had to go through a phase of epistemic concern and become an object of Eurocentric knowledge. This crisis-laden history is long, and really accelerated in the middle of the nineteenth century by the European discovery of the 'Ice Ages'. From the 1830s onwards, a Euro-Western fascination with glaciers and past evidence of ice-free landscapes elicits concerns that ice is not only changing, but changing as an effect of human actions, rather than natural variations in the earth's climate. How solid is it? It cracks and is heavily pressured, argued John Tyndall back in the 1860s. Others thought—more accurately, as it turns out—that it is a viscose substance. What no one thought of doing was asking indigenous and local communities who lived, worked and thought about ice what they thought. Nineteenth-century debates in Europe—rather than, say, indigenous communities in Peru or Mongolia—opened up ideas of deforestation affecting climate—and hence ice and snow. Svante Arrhenius' seminal 1896 paper established the connection between atmospheric CO_2 and global temperature, proposing that ice diminishes because of anthropogenic emission of greenhouse gases. His interest, however, was not deglaciation; on the contrary, he wanted to understand the causes of ice ages—at the time a prime concern in explaining the rich natural resources in Sweden, a country born out of 'ice and snow' (Dodds and Sörlin 2022). Besides, Arrhenius flatly denied the possibility that humans could burn enough carbon to seriously impact the climate.

Such optimism that human burning could not alter climate disintegrated well and truly some 50 years later. In the 1950s, the first projections came of an ice-free Arctic Ocean and large-scale melting of the ice sheets of Greenland and Antarctica (Siple 1953). It coincided with the early rise of 'the environment' as a crisis concept, identifying an ever-growing number of earthly features subjected to anthropogenic change. Ice—10 per cent of the total global terrestrial area—was gradually acknowledged as one of the parts that changed most. It was a theme in the International Geophysical Year (1957–1958)—the largest ever scientific enterprise—akin to a 'scientific Olympics' with 67 countries participating in research into the earth and heavens and 12 countries dedicating considerable resources to investigating the subterranean worlds of the Antarctic ice sheet. In the following decades, computer-based climate models became available and ice cores demonstrated drastic changes in global temperatures in the deep past (Edwards 2010). Ice was drawn into the growing data modelling complex of climate change science, and employed as a method for scientific visualisation.

It was in the aftermath of these insights—at a time when global interest had focused on the environment through the UN conference in Stockholm 1972, and climate was again on the agenda—that ice, or the cryosphere, became an environmental object, subject to concern and consideration, just like other environmental objects (e.g. atmosphere, oceans, deserts, biodiversity). We might say that it happened when Wally Broecker used the Dansgaard ice core to warn against his self-coined concept: 'global warming' (Broecker 1975). Changes in ice became an anticipated feature of each of the Intergovernmental Panel on Climate Change's assessment reports as of 1990. Their chronologies and trajectories as environmental objects differ, although they share the overarching understanding that there is 'the environment' out there that growing portions of humankind dangerously engage, extract and exploit. But for many decades still, concern over how glacial change affected local and indigenous communities in Greenland, the Andes or the melting Siberian tundra north was kept in the margins.

Privileged communities in the Global North have shifted from thinking of ice as simply an archive or a library to something that challenges our routinised application of frameworks and understanding. Ice has been, and continues to be, complicit with colonialism and extractive projects. We might even speak of a 'settler cryosphere'; understanding 'icy geopolitics' as an operating system that has probed ice, removed ice and assaulted the cold and darkness—and recognising that operating system as part of a continuum of activities that cherishes access, mobility and profitability. In Jarius Grove's terms, for centuries ice and indigenous peoples have been framed as part of 'savage ecologies' that need taming and improving and/or simply ignoring (Grove 2019). The nineteenth-century ice trade, for example, saw ice from the US end up in British India—acting as accomplice to colonial/tropical medicine and racialised tropes about the health of colonial authority. Imported ice had a distorting effect on local ice markets. Thus, an irony: ice was seen as obstacle to settler-colonial progress at the same time it was extracted and exported as part of a thriving international economy (bringing ice from the Americas to the British colonisers in India). And nowadays, we have artists and scientists extracting ice from the Arctic and bringing it to the metropoles. Arctic Base Camp, for example, transported a block of Greenlandic ice to the 2021 COP26 (26th Conference of Parties, informally known as the UN Climate Change Conference) meeting in Glasgow for the explicit purpose of staging an encounter between melting ice and climate change conference attendees (Figs. 2 and 3).

Part of the urgent task of ice humanities is to further scrutinise those stories, experiences and legacies that are irreducible to data points or objects passively awaiting extraction, disappearance and auditing by Euro-Western scientific bodies.

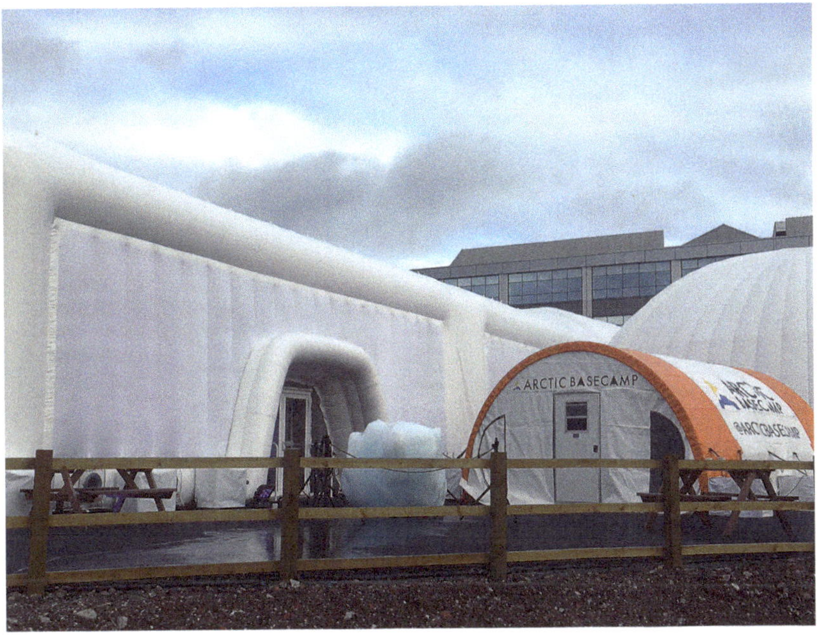

Fig. 2 Jelter Meers, The Arctic Base Camp at COP26 Glasgow, November 2021 © Jelter Meers

Future Ice Staging

What is our future relationship with ice? So far, it has been an explorer's nightmare. A scientific enigma. A source of myth and intrigue. A hotspot for tourism. An obstacle to resource extraction. A symbol of climate change and, now, crisis. An integral element in indigenous ways of living and engagement with the non-human world. Ice has been caught up in imperial imaginaries, informed settler-colonial fantasies, produced frontier masculinities and invited a range of aesthetic, cultural and mediated affects and responses.

Artists, writers, filmmakers, scientists and communities have been working with ice to communicate, engage and mobilise. Identities and subject-positions continue to be extracted and harvested from ice and in turn help to shape spaces for action as well as cultivate ambitions, desires and expectations. The popular entertainment industry in the UK continues to find opportunities to stage their own historical reconstructions

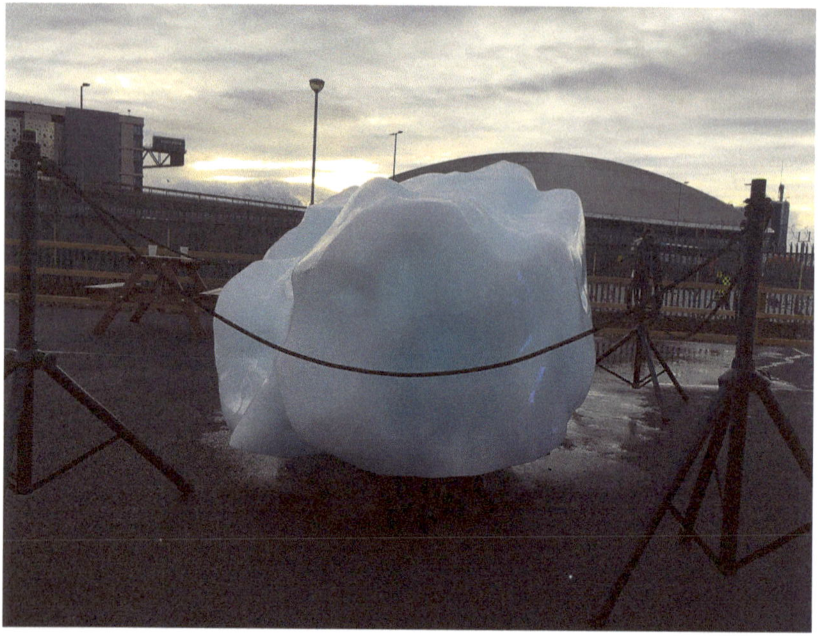

Fig. 3 Jelter Meers, The Greenlandic Ice Block by Arctic Base Camp, COP26 Glasgow November 2021 © Jelter Meers

(*North Water*), depicting a cold, terrifying Arctic attracting nineteenth-century adventurers and chancers. The Arctic continues to exercise a powerful cultural pull for those looking to stage the Arctic—and its ice—as exceptional.

Ice stages and staging ice is not therefore an invitation to turn the stage into a lazy metaphor or a leap into a pool of cynicism. Stage management carries with it the aura of fakery, manipulation and insincerity. Rather, it is an invitation to think about how engagement with ice is performative. We might point to a host of situated performances with ice. From the hosting of ice at major international conferences to glacial funerals, speeches, performances, treks and other actions. The performative politicisation of ice in indigenous struggle and environmental activism—although not without accusations of 'green colonialism' and external environmental activist insensitivities—foregrounds the antagonistic and contentious. Ice melt and disappearance will be understood and felt in uneven ways. There are deeply unequal socio-ecological capabilities and positionalities.

One way to negotiate the ice as stage is to think of it as encounter. The short film *Utuqaq* (2021), made by the cinematographer Iva Radiovjevic, premiered at the Camden International Film Festival, was released online (appropriately enough) on Earth Day and was shown elsewhere to critical claim. The film considers the intersection between four researchers working in the Greenlandic interior. They are performing the same kind of ice coring work that was pioneered during the Cold War years, when multinational groups of scientists and military personnel were eager to learn more about the island's climatological history and prevailing environment. This time, however, the focus is on a small English-speaking party while the commentary is entirely delivered in West Greenlandic by Aviaja Lyberth. The voiceover commentary and multi-perspective approach—combining sweeping landscape shots with more intimate explorations of scientific life at the base camp and immediate environs—achieves something quite notable. The viewer is invited to think about what is at stake as the researchers try to make sense of what is melting, disappearing and revealing. As the commentary notes, "The visitor knows that the borders of this landscape are melting ... what they don't know is what happens here". Footage of the scientists extracting, measuring and storing ice is interspersed with further commentary reflecting on the memory of ice, the harshness of the prevailing weather and the struggles to carry out even fundamental tasks (such as extracting ice cores from the machine itself). The "ice that lasts year after year" (*Utuqaq*) is in material crisis, and this film's narrative arc is ultimately one that invites us to think about ice as a protagonist, rather than a passive stage. As the film commentary asks: "What do [the scientists] want?". More poignantly, perhaps, the melting of Greenland reminds us that some worlds end up disappearing as a price for enabling the continued existence of other worlds.

Acknowledgements I am indebted to the editors for their encouragement and to Professor Harriet Hawkins for her supportive comments on an earlier draft.

Bibliography

Alley, Richard. 2002. *The Two-Mile Time Machine: Ice Cores, Abrupt Climate Change, and Our Future*. Princeton: Princeton University Press.

Antonello, Alessandro, and Mark Carey. 2017. Ice Cores and the Temporalities of the Global Environment. *Environmental Humanities* 9: 181–203. https://doi.org/10.1215/22011919-4215202.

BBC. 2019. The Dying of the Ice. *Radio 4*. https://www.bbc.co.uk/programmes/m0006mqw

Bloom, Lisa. 1993. *Gender on the Ice*. Minneapolis: University of Minnesota Press.

———. 2022. *Climate Change and the Polar Aesthetics: Artists Reimagine the Arctic and Antarctic*. Durham/London: Duke Press.

Bloom, Lisa, Elena Glasberg, and Laura Kay. 2008. Gender on Ice. *S & F Online*. https://sfonline.barnard.edu/ice/bloom_02.htm

Broecker, Wallace. 1975. Climatic Change: Are We on the Brink of a Pronounced Global Warming? *Science* 189: 460–463. https://doi.org/10.1126/science.189.4201.460.

Carey, Mark. 2010. *In the Shadow of Melting Glaciers in the Shadow of Melting Glaciers: Climate Change and Andean Society*. Oxford: Oxford University Press.

Cosgrove, Denis, and Veronica Della Dora. 2008. *High Places*. London: I. B. Tauris.

Craciun, Adriana. 2016. *Writing Arctic Disaster*. Cambridge: Cambridge University Press.

Cruikshank, Julie. 2001. Glaciers and Climate Change: Perspectives from Oral Tradition. *Arctic* 54 (4): 377–393. https://doi.org/10.14430/arctic795.

———. 2005. *Do Glaciers Listen? Local Knowledge, Colonial Encounters, and Social Imagination*. Vancouver: UBC Press.

Dodds, Klaus. 2018. *Ice: Nature and Culture*. London: Reaktion.

———. 2021. Geopolitics and Ice Humanities: Elemental, Metaphorical and Volumetric Reverberations. *Geopolitics* 26: 1121–1149. https://doi.org/10.1080/14650045.2019.1697240.

Dodds, Klaus, and Jen Rose Smith. 2022. Against Decline? The Geographies and Temporalities of the Arctic Cryosphere. *Geographical Journal*. https://doi.org/10.1111/geoj.12481.

Dodds, Klaus, and Sverker Sörlin. 2022. *Ice Humanities*. Manchester: Manchester University Press.

Doel, Ron, Kristine Harper, and Matthias Heymann. 2016. *Exploring Greenland: Cold War Science and Technology on Ice*. London: Palgrave.

Edwards, Paul. 2010. *A Vast Machine: Computer Models, Climate Data, and the Politics of Global Warming*. Boston: MIT Press.

Elmore, Aurora, Nicole Alexiev, and Valerie Craig. 2020. Understanding the World's Water Towers Through High-Mountain Expeditions and Scientific Discovery. *One Earth* 3: 561–563. https://doi.org/10.1016/j.oneear.2020.10.017.

ESA. 2021. Antarctic Glacier Named Glasgow to Make COP26. Website. https://www.esa.int/Applications/Observing_the_Earth/Space_for_our_climate/Antarctic_glacier_named_Glasgow_to_mark_COP26. Accessed 10 Nov 2022.

Farish, Matt. 2010. *The Contours of the Cold War*. Minneapolis: University of Minnesota Press.

Gagné, Karin. 2019. *Caring for Glaciers: Land, Animals, and Humanity in the Himalayas*. Seattle: University of Washington Press.

Gearheard, Shari, Lene Holm, Henry Huntington, Joe Leavitt, and Andrew Mahoney. 2013. *The Meaning of Ice: People and Sea Ice in Three Arctic Communities*. Hanover: International Polar Institute Press.

Grove, Jarius. 2019. *Savage Ecology*. Durham: Duke University Press.

Hastrup, Kirsten, and Karen Olwig. 2012. *Climate Change and Human Mobility*. Cambridge: Cambridge University Press.

Herzberg, Julia, Christan Kehrt, and Franziska Torma. 2018. *Ice and Snow in the Cold War: Histories of Extreme Climatic Environments*. Oxford: Berghahn Books.

Hosseini, Anita. 2019. Olafur Eliasson's 'Ice Watch'. *Bilderfahrzeuge Blog*. https://bilderfahrzeuge.hypotheses.org/3528

Jackson, M. 2019. *The Secret Lives of Glaciers*. Brattleboro: Green Writers Press.

Jones, Max. 2004. *Last Great Quest*. Oxford: Oxford University Press.

Leane, Elizabeth. 2020. *South Pole*. London: Reaktion.

Lincoln, Amber, Jago Cooper, and Jan Loovers. 2020. *Arctic: Culture and Climate*. London: British Museum Publishing.

Mancilla, Alejandra. 2021. From Sovereignty to Guardianship in Ecoregions. *Journal of Applied Philosophy*. https://doi.org/10.1111/japp.12561.

O'Reilly, Jessica. 2017. *The Technocratic Antarctic*. Cornell: Cornell University Press.

Radiovjevic, Iva. 2021. *Utuqaq*, 27 minutes.

Roberts, Peder. 2011. *The European Antarctic: Science and Strategy in Scandinavia and the British Empire*. London: Palgrave.

Rockström, Johan. 2009. Planetary Boundaries: Exploring the Safe Operating Space for Humanity. *Nature* 461: 472–475.

Ruiz, Rafico, Paula Schönach, and Rob Shields. 2020. Special Issue. Beyond Melt. Indigenous Lifeways in a Fading Cryosphere. *Northern Review* 13: 7–17.

Schmidt, Jeremy. 2021. Glacial Deaths, Geologic Extinction. *Environmental Humanities* 13: 281–300.

Siple, Paul. 1953. *Proposal for Consideration by the US National Committee (UGY)*. 1 May 1953, C1, USNC-IGY. Washington, DC: National Academy of Sciences.

Tall Bear, Kim. 2017. Beyond the Life/Not Life Binary: A Feminist–Indigenous Reading of Cryopreservation, Interspecies Thinking, and the New Materialisms. In *Cryopolitics: Frozen Life in a Melting World*, ed. Joanna Radin and Emma Kowal, 179–202. Cambridge, MA: MIT Press.

Watt-Cloutier, Siila. 2015. *The Right to Be Cold: One Woman's Story of Protecting Her Culture, the Arctic and the Whole Planet*. Toronto: Penguin Canada.

Zhang, Fan, Sahadeep Thapa, Walter Immerzeel, Hongbo Zhang and Arthur Lutz. 2019. Water Availability on the Third Pole. *Water Security* 7 Article 100033.

Movies on Ice: An ArtSci Perspective on Communicating Antarctic Ice in the Climate Emergency

Craig Stevens and Gabby O'Connor

Climate Emergency and the Next Generation

Our species and its cultures are at a crossroads in terms of the choices and outcomes related to greenhouse gas emissions and global heating, including the polar ice caps being entrained by, and into, the warming ocean. The impacts will extend far beyond sea level rise. Geophysical, ecological, political, economic, social and cultural consequences, some of which cannot be readily anticipated. How fast they will happen—decades and centuries into the future—depends on decisions made now and in the coming

C. Stevens (✉)
National Institute of Water and Atmospheric Research, Wellington, New Zealand

University of Auckland, Auckland, New Zealand
e-mail: Craig.Stevens@niwa.co.nz

G. O'Connor
University of Auckland, Auckland, New Zealand

A. Hemkendreis, A.-S. Jürgens (eds.), *Communicating Ice through Popular Art and Aesthetics*, Palgrave Studies in Media and Environmental Communication,
https://doi.org/10.1007/978-3-031-39787-5_3

decade. Projections of surface temperature trajectories over the next 15 years suggest there is not a great deal to distinguish the acceptable outcomes from the poor-to-appalling trajectories (Boulton 2022). The topic is so pervasive that most audiences have 'made up their mind' about the issue, which means that even very clear messaging is unlikely to suddenly win people over regardless of their position. Thus, there is a significant potential benefit in diversifying how we communicate scientific knowledge about future climates and impacts.

Responses to climate are happening via a range of pathways—from the global multi-government IPCC (Intergovernmental Panel on Climate Change) initiative (Lawrence et al. 2022) through to individual protests (Murphy 2021). Art is another such way of engaging with, understanding and interpreting our environment that can sometimes sidestep audience reticence. This chapter makes connections between science, art, climate and popular culture using educational, artistic and communication threads that have emerged in the long-standing collaboration between Antarctic scientists (including Stevens) and an artist (O'Connor). We first describe the nature of our art-science collaboration, which included: joint public and educational talks; O'Connor participating in science fieldwork and creating scientific data; scientists helping in (co-)creating art and activities, to enhance the connection between art and wider understandings of climate. We then use themes that emerge from popular movies both as a way of humanising climate research and as a gateway to representing ice and climate in art and science. Finally, in the discussion, we consider several questions that emerge from our own experiences: (a) What are the creative, aesthetic, visual art and socio-cultural results of science-art collaboration, and can science-based artistic communication be defined? (b) Can an aesthetic experience communicate research on ice and its crucial importance for our planet? (c) What role does science play in, and for, our socio-cultural ideas or imaginaries of ice?

Art-Science Collaboration

The collaboration central to this paper focused on developing an approach to communicating climate science by engaging young people in the art-making process. The initiative seeks to influence how the next generation view their choices, both in terms of education and the climate emergency. It results in cultural validation of outputs created for public exhibitions. This was built on a foundation of having the artist as an equal partner in the boots-on-ice science and conceiving of creativity as equally spread throughout the art-sciences. One of us (O'Connor) was embedded in the science—by participating in several Antarctic science 'field events'—while at the same time developing her own art practice.

The ideas and thinking inherent in research need a genesis. O'Connor's initial artworks were inspired by views out airplane windows flying over the ice-covered Arctic Ocean, as well as the cultural perspectives on the fate of the unretrieved remains of Scott and colleagues, who perished on the Ross Ice Shelf in the early 1900s (O'Connor 2011, Fig. 1). Objects frozen within ice shelves move based on a complex balance of the flow and stretching of the ice shelf itself (which can reach speeds of several metres a day), snow accreting from above and the ice melting or re-freezing from below. The idea that these explorers are posthumously continuing their journey, now with the aid of geophysical mechanics, inspired O'Connor's

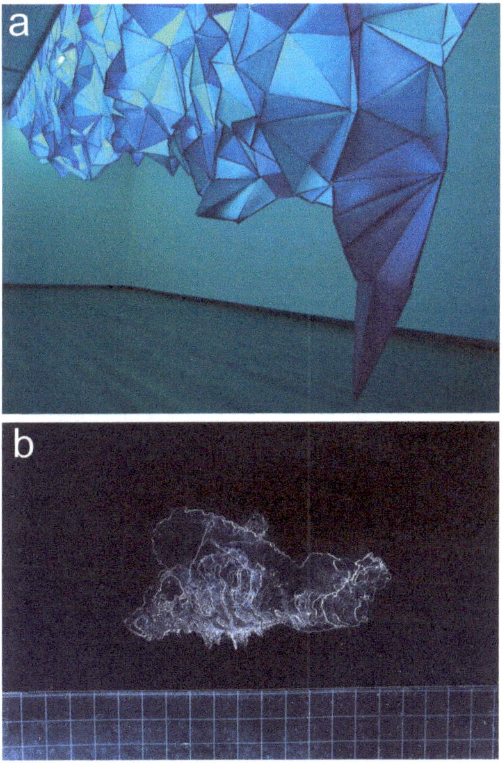

Fig. 1 Different scales of ice-focused art, (**a**) Gabby O' Connor, the paper iceberg installation from What Lies Beneath, 2011, (**b**) a photograph of an ice crystal produced for both art and science, both © Gabby O' Connor

What Lies Beneath (Fig. 1a) installation. This work consisted of a large three-dimensional representation of an ice shelf constructed from triangular panels of lacquered tissue paper dyed various shades of blue. While the work did not figuratively include the explorers, it did invoke their resting place within the three-dimensionality of an ice shelf viewed from below.

Publics often assign substantial value to the act of going to Antarctica. In our collaboration, the artist O'Connor prepared material for scientists to deploy in Antarctica (Stevens and O'Connor 2016), which led to her participating in research expeditions (O'Connor and Stevens 2018). Crucially, O'Connor contributed to the science by making detailed measurements of individual ice crystals, something that had never been done before. This involved developing a sampling, photographic and measurement protocol that extracted crystals from the underside of the sea ice (using a specially developed scoop) and then quickly moved them to the photographic stage before they melted or re-froze (Stevens et al. 2019). In addition, she documented both the art and science in unique ways, not normally captured by the science team's processes; whereas the science team typically produced a time-stamped diary, O'Connor augmented this with videos, photographs and sketches of the day-to-day activity.

There is a substantial historical context for the intersection of art and science in an Antarctic setting (Stevens et al. 2019), with a range of impacts, from science and environmental communication to the values associated with art. Our collaboration also developed educational activities designed to engage young audiences with science, art and climate (O'Connor and Stevens 2015), which has had a significant impact (Lawrence et al. 2022). We developed workshops for school-aged students that simulated aspects of fieldwork, helped them make artworks and gave climate science social license. The initial school workshops were primarily about communicating climate science and the nature of creativity, although several additional themes quickly emerged, such as how using art as a gateway to science democratises science and learning. Some students think of themselves as not knowing anything about science, but they will happily help build a glacier artwork. With suitable tutoring, as they create they learn through discussion about ice mechanics, computer modelling, teamwork and oceanography. This of course requires skilled and knowledgeable tutors willing to present the science as accessible.

When looking at central themes for the future of our species like climate and ecosystem resilience, where long-term systems need to be supported by immediate and widely supported decisions, it is clear that multi-generational perspectives are required. While the workshop outcomes were accessible to all ages, the approach worked best as a Trojan Horse for communicating concepts to family and social groups, by positioning young people as the vector for climate-futures awareness. The school students worked on their art-science project and then, once it was on display in a gallery, they proudly talked to their family about what they knew about Antarctica and climate. This gives the initiative longevity well beyond the typical exhibition.

When talking to audiences about Antarctica and its place in our climate system we can use these science and art experiences productively in several ways. Centrally, we can show a wide variety of data plots, photographs of scientific equipment, computer simulations or artworks produced in, or about, Antarctica. Beyond this, it also helps to consider the audience's understanding of Antarctica and climate, as movies are one of the primary ways people connect with an environment they will probably never see or experience. Even for people who do get to experience these environments—like Polar researchers—movies provide enjoyment, imagination, entertainment and learning, making them a shared entry-point into valuing our climate system.

Movies on Ice

In this section we humanise the people who collect some of the key pieces of information being used to project what future climate and sea levels will be. An Antarctic field camp is where (some) scientific data are collected, and where researchers embed themselves in an environment that allows them to make mental notes that are otherwise difficult to achieve from a desk—even if the latter is better connected to the wider datasets necessary for climate science. As well as a desire to discover new ice knowledge, such field camps are fuelled by collaborative work practises and socialisation. These are often built around shared popular culture—something that goes back through the history of Polar exploration. For example, the bar at Scott Base (the New Zealand research base on Ross Island, 77°51′S 166°46′E) used to have a wall-sized photograph showing Scott's party enjoying a mid-winter meal, crowded around a long table; expeditioners found ways of amusing themselves with theatre, newspapers, crafts, art and games through the long Polar winter (Pearson 2004). Things are a little different now that expeditions are so much shorter, and we can travel with hard drives of entertainment.

The first Antarctic field season for Stevens took place in 2004 at a sea ice camp in southern McMurdo Sound (77° 45′ S 166° 5′ E)—a suite of converted insulated shipping containers lined up on sea ice floating on an ocean half a kilometre deep (Fig. 2). Several of the containers had hatches in their floors through which 'hydroholes' were cut to access the ocean below. The field team worked in these for several weeks, sleeping in one container, cooking and eating in another and sampling from the hydroholes in another. The 24-hour daylight and ready access to the ocean make it easy to overwork so there are deliberate strategies to take a break; one of which is 'movie night on ice'. We would set up the biggest screen and best sound we had and sit together to watch movies. This collective activity echoes early expeditions watching shared entertainment to pass the time and bring the team closer.

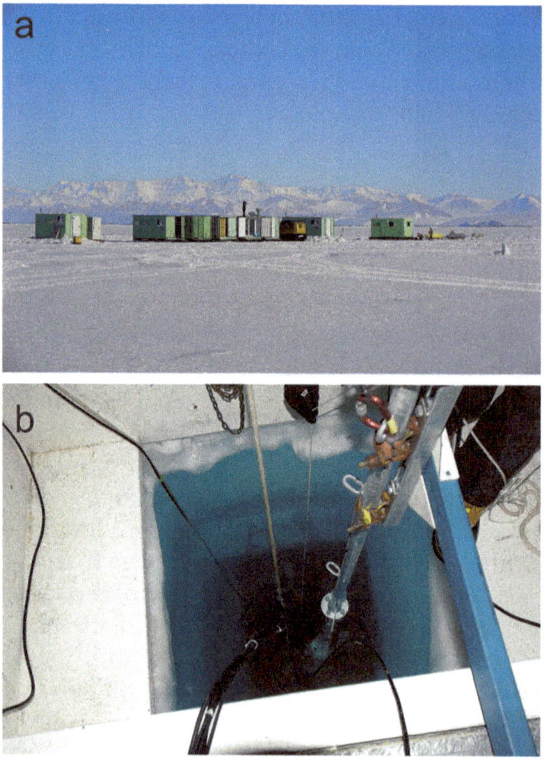

Fig. 2 Container-based sea ice field camp, (**a**) B. Grant and NIWA, the sea ice is two metres thick, (**b**) several of the containers have holes in their floor and through the ice © Stevens/NIWA

The first movie watched while floating on ice was *The Day After Tomorrow* (Emmerich 2004), a story of family resilience in a sudden icy climate emergency. Generally, ice-based movies were the last thing people in the team wanted to watch; they were living on the biggest lump of ice on the planet and did not really need more—but there were exceptions if the movie was spectacular, scary or so perfectly aligned (like *The Day After Tomorrow*) that it had to be watched. In the wider world, the success of the movie was used as an opportunity to communicate climate science (Hansen et al. 2004; Hart and Leiserowitz 2009) although the science presented was criticised as unrealistic and alarmist (Bilandzic and Sukalla 2019). However, it is almost certain the critics did not watch it while sitting in a container floating on ice dozens of kilometres from the nearest help. The movie opens with a field science team in an Antarctic ice shelf camp just as the ice shelf cracks apart revealing a yawing chasm. Our team were on sea ice (typically only a few metres thick) and so very different to an ice shelf (sometimes a kilometre thick). If sea ice cracks apart the ocean is only a few tens of cm away, but it is cold, and the cracking would likely mean we were adrift. Consequently, the distinction did not seem so great to the team at the time. For all its forward projection around climate and poking a stick at contemporary government policies, the movie story is a conservative one about a dad—contrasted with his caring doctor ex-wife—trying to be a better man after having given too much of himself to science. This partly resonated with our Antarctic audience, as field science takes people away from their families for long periods. There were some evocative scenes in the movie relating to weather and extreme cold and how humanity will cope. An icicle-laden Statue of Liberty used as a key image was both visually arresting and echoed the final scene in *The Planet of the Apes* (Schaffner 1968), where the crash-landed space-travelling protagonist escapes captivity only to find the Statue of Liberty submerged in sand, implying he had time travelled to a post-nuclear Earth (Kirshner 2001). Time travel is often invoked in climate understanding, as our greenhouse gas emissions take the planet back to a time when sea levels, temperatures and weather were vastly different.

As well as the symbolism of the fallen Statue of Liberty, *The Day After Tomorrow* and *The Planet of the Apes* share a sense that our way of life may wreck what we hold dear. At the time *The Day After Tomorrow* was released, the public questioned its veracity and whether climate change is even real. Perhaps climate movies needed to shift further from reality as—despite the far-fetched nature of *Planet of the Apes*—one comes away with a clear sense that nuclear Armageddon will not have good outcomes for humanity. If you have recently escaped floodwaters, you are unlikely to quibble about the finer science details of a movie that suggests flooding is not good. For example, Selvey et al. (2022) describe how Queensland voters can ignore the evidence of climate change (based on data from 2019). However, there is clearly some tipping point, as only a few months after this was published, a Greens Party candidate won the electorate in question (Brisbane, Australia); the region had just experienced its worst recorded flooding.

Getting on with people is a critical part of working in Antarctica. Ice camps the authors have experienced had anywhere from 5 to 25 people living and working in them and were isolated from the large bases; it is far more efficient to be a small group with focused tasks in the right location than to commute from a large base. Typically, these small groups are drawn from a narrow demographic. We are not out in the field for very long—it is too expensive these days. However, the close work and shared experiences make for connections not found 'back in the world'—a phrase from leading ice drilling engineer Darcy Mandeno, who has spent more time than most in ice camps.

John Carpenter's *The Thing* (1982) explores the nature of group dynamics in an extreme way that makes for a perfect movie night classic, especially when watched in Antarctica. The movie opens with a bang: A rogue husky, infected with an alien virus, enters an Antarctic field base, is chased and shot at by a helicopter which subsequently crashes. The 12-person base team is all-male, not uncommon at the time, and with the usual set of character tropes. The researchers slowly become aware of what they are up against as they get picked off by the virus, one by one, in glutinous non-CGI detail. The dwindling pool of survivors tries to work out what is going on and who is, and is not, infected. The theme is unfortunately just right for this worldwide pandemic era and gives recent viewers a new perspective. The movie was released in 1982, a decade prior to the cessation of using dog teams in Antarctica due to biosecurity; there was concern for disease transmission and disturbance to seals and penguins—a theme that parallels the fears of a contagious biological mayhem unleashed in Carpenter's movie. The ending leaves the viewer uncertain as to the ultimate fate of the two survivors, but they have an unspoken collective resolve.

The composer Ennio Morricone created a theme within the score of *The Thing* that is seemingly based around the human heartbeat. Snow is a good absorber of sound so standing still on snow on a windless day you can hear your own heartbeat. This is only if you are far enough away from the generators. When within proximity to the camp generators, there is a constant hum that underpins all activities. Once one becomes accustomed to this constant hum sound becomes a sensory input in a place where this is limited. The protected containers of the camp (Fig. 2) and long contemplative hours of sampling are well-suited to having music on—either quietly or (more frequently) loudly, depending on the mood. These are not the only sounds, though. The not-uncommon storms whistle past the containers, rattling and moaning. At night, lying in either a freezing-cold lower bunk or boiling-hot upper bunk, it is possible to hear seals calling across their ocean domain just a few metres below. The Antarctic soundscape ducks, weaves, clicks and vibrates, as an other-worldly parallel to Carpenter's 1970s synthesis technology of oscillators, sweeping filters and sequencers.

One night, in the wrap-up phase of a recent expedition, 22 scientists and engineers living and working on a remote corner of the Ross Ice Shelf (82° 28′S, 155° 17′W)—sitting on 500 metres of ice, 1000 kilometres from the nearest base—crammed into the mess tent to watch the recently released 2021 filmic interpretation of Frank Herbert's novel *Dune*. The movie is quite long, and bedtimes are typically early so we were fortunate that the generator needed re-fuelling at roughly the mid-point, giving us an excuse to split the movie over two nights. Unfortunately, because of the 24-hour daylight of the high latitude summer, the darker scenes in the film were not projected well (Fig. 3). On the other hand, the sound was fantastic (even with the generator), and effectively immersed us in the sense of a planet covered in sand where water is scarce enough that people wear suits to recycle all fluids. Given that we were living in a desert and were assigned "P bottles"—so that we are not randomly generating yellow patches of snow everywhere—the story did not feel that abstract. The worlds of *Dune*—Caladan, Arrakis and others—are clearly recognisable as perturbations of our own climate system. While the connection between life on these earth-ish planets and our own future is reasonably apparent (Buse 2020), it would have been less so in the 1960s when Herbert wrote his novel. There was real cognitive dissonance sitting in that tent and being emotionally affected by a barely visible artistic representation of climate future in a setting that was equally strange—but real and connected to the transition from one state to the other.

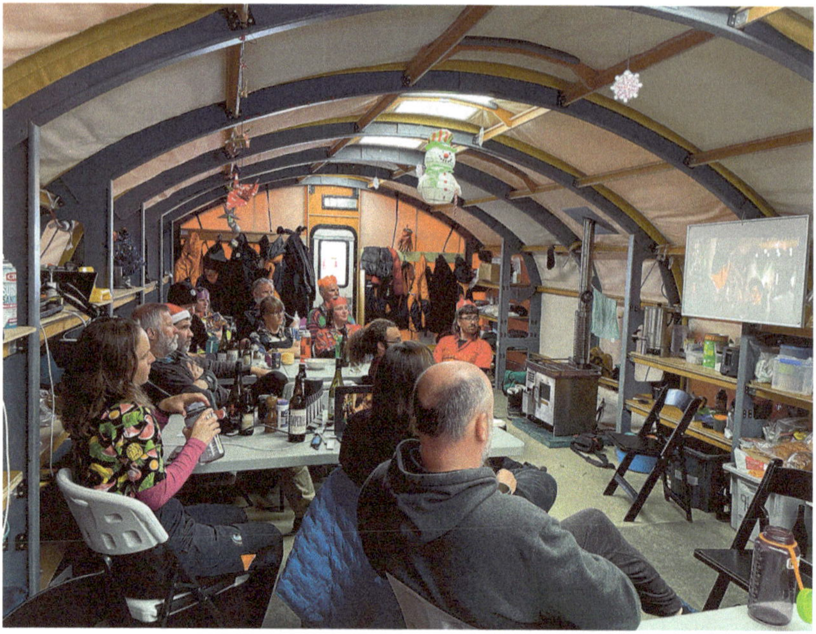

Fig. 3 Movie night in the mess tent with 24-hour daylight, © Stevens

Less than two decades after the release of *The Day After Tomorrow*, as climate-driven disasters become more and more common, the planet is changing in both predictable and unpredictable ways. While the slow trends in things like ocean heat content and tropical storm frequency are of central concern as they are pathways to the desert of Arrakis in *Dune*, so too are the sudden localised disasters dramatised in *The Day After Tomorrow*. Powerful visual and aesthetic representations of change are one way to engage publics and decision-makers beyond the known narrative.

Representing What Lies Beneath

Movies provide a pathway to represent both icy worlds and climate narratives but how can visual information be used to communicate climate research? In this section we discuss how a range of graphical representations help explain our Antarctic research. The importance of clear and meaningful representations of climate science can be seen in the work explaining infectious disease science to a population grappling with the COVID-19 pandemic. One of the many key messages is the beneficial impact of well-designed graphical representations of scientific concepts (Beattie and Priestley 2021). The best of these helped (and continue to help) both publics and decision-makers understand what was at stake and take actions to mitigate future risks; indeed, "flatten the curve" (Li and Molder 2021, 899) helped scientists understand how to make their work impactful. In terms of science communication, and especially compared to climate, COVID entered hyperspace (Rauchfleisch et al. 2021). After decades of climate scientists trying to get wider publics interested in forecasts—of things like air and ocean temperatures, flood probabilities and so on—suddenly, within a few months, public discourse and interpretation started to successfully tackle complex quantitative descriptions of recent outcomes, future scenario-based projections and probability distributions through science communication about COVID.

Most people are familiar with various climate change diagrams, including the 'hockey stick graph', showing dramatically rising values of sea level, or atmospheric temperature, or bushfires, or floods (Mann 2021). Graphical representation in science is only now emerging from the era when data appeared in 'plots' and plots appeared in 'papers', on paper. Until recently, figures in colour or photographs cost authors extra. It is still the case that animations are not embedded in papers. Scientists retiring now would have started their careers with talks based around overhead projector transparencies, or even with line diagrams photographically transferred to 35 mm slides with a dark blue background and white lines. In contrast, scientists starting out now can distil their research topic into a 30 second TikTok video and then instantly distribute it to many thousands of people.

Whether climate-centred scientific diagrams are more-traditional black-and-white graphs or more-modern virtual reality visualisations of data, they all tell the same story of dramatic change. Our own science-art collaboration has influenced how we present the science and art—and their intersection. As mentioned, artist O'Connor was involved in the production of scientific data that then found itself in science papers (Stevens et al. 2019). Stevens was inspired to incorporate more cartoon-style graphics into scientific presentations to help explain observations and conclusions—not for frivolity but for simplifying a message from a wall of complexity, jargon and seriousness (Jonsson and Grafström 2021). One of the most effective methods we have found for communicating about our recent expeditions—to both public and research audiences—is short video-format sequences showing underwater scenes or field teams working (see supplementary video in O'Connor and Stevens 2018). Indeed, short movies and timelapses of art production and trans-disciplinary fieldwork impact pedagogical settings and exhibitions through their attention-grabbing, informative potential and their ability to invite a suspension of disbelief that is vital to making significant changes to ways of thinking and working. The relative ease with which we can capture high-quality video imagery of the ocean beneath ice enables us to engage with media outlets and do voiceovers with audience-focused messages. While such views are far removed from most audience experiences and have visual and aesthetic value, they are still an exploration of scientific data. Perhaps most importantly, they feed into evolving approaches to improved communication both using art and science, separately and together.

By taking these art-data and modulating them with experience of the environment, the artist develops pathways to science that do not start with a prescriptive 'this way to being told about science' (Menezes et al. 2022). What emerges ranges from photographic representations—equally appropriate for a gallery or a scientific journal—through to works built from completely unrelated material that represent intangible cultural tropes or scientific elements, such as connectivity and cross-cultural values (Hüppauf and Weingart 2007).

Discussion

What are the aesthetic, visual art and socio-cultural outcomes, and can science-based aesthetic communication be defined? Presentation and communication of scientific data have some well-established rules and structures, yet recently these approaches have started undergoing dramatic changes. Conversely, art has always sought to challenge rules and break down structures. O'Connor's artistic practice—emerging from on-ice experiences—is multi-faceted and reflects the changes in science communication and inherent artistic deconstruction. O'Connor produced ice sculptures prior to the field work that morphed into representations of the crystals coating the underside of sea ice. In a gallery setting, this was arranged as a ceiling above a mat that the audience was invited to lie upon and look up into the ice crystal boundary—highly evocative of the fieldwork that produced underwater imagery without being a direct representation. In addition, photographs of individual ice crystals were displayed in large formats, like a portrait gallery, which imbued them with their own personality. Other works included A3-sized watercolours of both small- and large-scale scenes with a dramatic palette of pinks and blues. Finally, video loops of the ice crystals on a rotating Lazy Susan stage invoked the research process by showing how research evolves—in this case from using crystals made from clear plastic tape through to the final imagery of real crystals. This multiplicity of formats increases the probability of finding a way to make any given audience member expand their perspective (Cozen 2013). O'Connor refers to 'hijacking' the traditional science talk by incorporating visual art sensibilities—which, in turn, inspires scientists to improve how they communicate their science (Fig. 4).

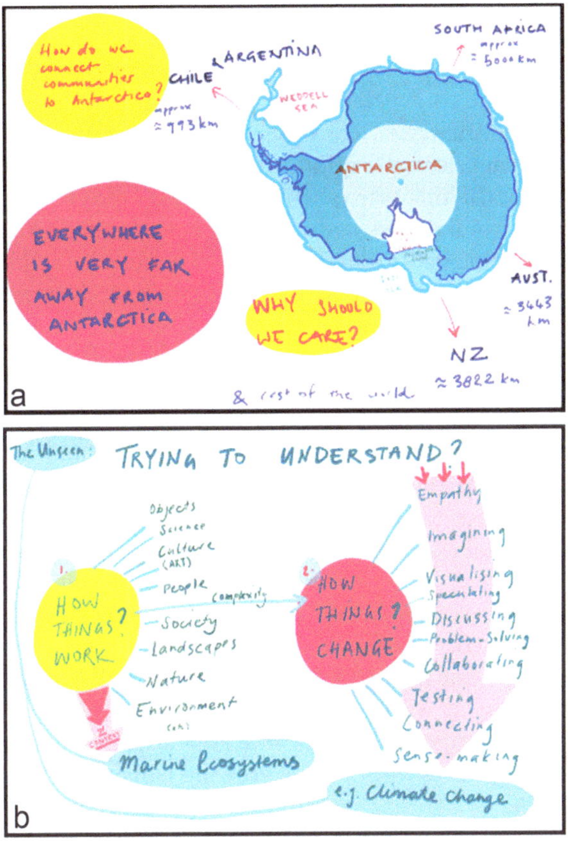

Fig. 4 Diagrams for talks presented to scientists and social scientists given by an artist, © Gabby O'Connor

Can art communicate research on ice and its crucial importance for our planet? There is a long history of artists working in Antarctica (Jackson 2019; Stevens et al. 2019) although never in great numbers. While the resulting works are art in themselves with genuine aesthetic value, they enhance the science communication by providing a different—but related, and equally valid—perspective on ice and climate science. Art can provide a personal, humanised view or one that distils beauty or harshness. It is a view unencumbered by graphs and numbers. Beyond this, what interests us is when people do not know they are learning from the art; a kind of bimodal Trojan Horse allows concepts from parallel streams to cross over and enhance conceptual awareness.

What role does science play in, and for, our cultural ideas or imaginaries of ice? There is likely a hemispherical division in answering this question. The Arctic—despite its history of exploration and apparent impenetrability—has been home to peoples for millennia; science is just one thread in the Arctic story. Antarctica, conversely, is imaginatively and realistically within the domain of the explorer and scientist, as mandated by the Antarctic Treaty. In this way, scientists—if not always the science—are central to our cultural understanding of Antarctic ice. Due to the relatively narrow demographic background of Antarctic researchers, this also means that we have a narrow perception of Antarctic ice. In just one dimension of diversity, Antarctic field operations are one of the last arenas to allow women to participate on anything close to equal standing, which leaves a lingering legacy that remains a challenge. The future of successful science requires diverse scientists who think broadly about interpretation, representation and impact (Bernard and Cooperdock 2018).

We speculate that the new generation of scientists are more open to disciplinary synergies and crosstalk than their academic forebears and that this will enable wider audiences, given sufficient context, to embrace the shared creativity and research structures of science. Our own ice-related trans-disciplinary research yielded tangible outcomes for climate science, art, education and communication, and gives us confidence that audiences can be trusted to understand science. These aspects appeal to different audiences in different ways but the hope is that people can come away with an expanded appreciation of the role Polar expanses of ice will increasingly play in all our lives through the growing climate challenges.

Acknowledgements We thank the many colleagues who have provided support and perspectives on this collaboration and the role of art in helping engage with Antarctica, including Natalie Robinson, Brett Grant, Tim Haskell, Craig Stewart, Christina Hulbe, Huw Horgan, Nick Golledge, Darcy Mandeno, Pete de Joux, Andy Falconer, Gavin Dunbar and Inga Smith. This work was made possible by a range of funding sources including the New Zealand Antarctic Research Institute, Marsden Fund, Deep South and Sustainable Seas National Science Challenges and the NZ Ministry for Business Innovation and Employment Antarctic Science Platform (ANTA1801). We thank Antarctica New Zealand for their support of the field operations.

Bibliography

Beattie, Alex, and Rebecca Priestley. 2021. Fighting COVID-19 with the Team of 5 Million: Aotearoa New Zealand Government Communication During the 2020 Lockdown. *Social Sciences & Humanities Open* 4 (1) 100209:1–10. https://doi.org/10.1016/j.ssaho.2021.100209.

Bernard, Rachel E., and Emily H. Cooperdock. 2018. No Progress on Diversity in 40 years. *Nature Geoscience* 11 (5): 292–295. https://doi.org/10.1038/s41561-018-0116-6.

Bilandzic, Helena, and Freya Sukalla. 2019. The Role of Fictional Film Exposure and Narrative Engagement for Personal Norms, Guilt and Intentions to Protect the Climate. *Environmental Communication* 13 (8): 1069–1086. https://doi.org/10.1080/17524032.2019.1575259.

Boulton, Chris A. 2022. Looking to the (Far) Future of Climate Projection. *Global Change Biology* 28 (2): 346–348. https://doi.org/10.1111/gcb.15936.

Buse, Katherine. 2020. The Working Planetologist: Speculative Worlds and the Practice of Climate Science. *Culture & Theory* 202: 51. https://doi.org/10.14361/9783839447512-003.

Carpenter, John. 1982. *The Thing*, Universal Pictures, 108 minutes.

Cozen, Brian. 2013. Mobilizing Artists: Green Patriot Posters, Visual Metaphors, and Climate Change Activism. *Environmental Communication: A Journal of Nature and Culture* 7 (2): 297–314. https://doi.org/10.1080/17524032.2013.777353.

Emmerich, Roland. 2004. *The Day After Tomorrow*, 20th Century Fox, 123 minutes.

Hansen, Bogi, Svein Østerhus, Detlef Quadfasel, and William Turrell. 2004. Already the Day After Tomorrow? *Science* 305 (5686): 953–954. https://doi.org/10.1126/science.1100085.

Hart, Philip S., and Anthony A. Leiserowitz. 2009. Finding the Teachable Moment: An Analysis of Information-Seeking Behavior on Global Warming Related Websites During the Release of The Day After Tomorrow.

Environmental Communication 3 (3): 355–366. https://doi.org/10.1080/17524030903265823.

Hüppauf, Bernd, and Peter Weingart. 2007. Images in and of Science. In *Science Images and Popular Images of the Sciences*, ed. Bernd Hüppauf and Peter Weingart, 3–31. New York: Taylor & Francis Group.

Jackson, Adele. 2019. A Changing Cultural Climate: Realising the Value of Artists Working in Antarctica. *Polar Record* 55 (5): 351–357. https://doi.org/10.1017/S0032247419000470.

Jonsson, Anna, and Maria Grafström. 2021. Rethinking Science Communication: Reflections on What Happens When Science Meets Comic Art. *JCOM* 20 (02): Y01:1–16. https://doi.org/10.22323/2.20020401.

Kirshner, Jonathan. 2001. Subverting the Cold War in the 1960s: Dr. Strangelove, the Manchurian Candidate, and The Planet of the Apes. *Film & History: An Interdisciplinary Journal of Film and Television Studies* 31 (2): 40–44.

Lawrence, Judy, Brendan Mackey, Francis Chiew, Marc Costello, Nina Lansbury, Kevin Hennessy, Uday Nidumolu, Gretta Pecl, Lauren Rickards, Nigel Tapper, Alistair Woodward, and Anita Wreford. 2022. Chapter 11: Australasia. In *Working Group II: Impacts, Adaptation and Vulnerability, Sixth Assessment Report*, ed. Hans-Otto Pörtner and Debra Roberts, 1–151. Cambridge: Cambridge University Press.

Li, Nan, and Amanda L. Molder. 2021. Can Scientists Use Simple Infographics to Convince? Effects of the "Flatten the Curve" Charts on Perceptions of and Behavioral Intentions Toward Social Distancing Measures During the COVID-19 Pandemic. *Public Understanding of Science* 30 (7): 898–912. https://doi.org/10.1177/09636625211038719.

Mann, Michael E. 2021. Beyond the Hockey Stick: Climate Lessons from the Common Era. *Proceedings of the National Academy of Sciences* 118E (39) p.e2112797118:1–9. https://doi.org/10.1073/pnas.2112797118.

Menezes, Sunshine, Kayon Murray-Johnson, Hollie Smith, Hannah Trautmann, and Mehri Azizi. 2022. Making Science Communication Inclusive: An Exploratory Study of Choices, Challenges and Change Mechanisms in the United States from an Emerging Movement. *Journal of Science Communication* 21 (5): A03:1–22. https://doi.org/10.22323/2.21050203.

Murphy, Patrick D. 2021. Speaking for the Youth, Speaking for the Planet: Greta Thunberg and the Representational Politics of Eco-celebrity. *Popular Communication* 19 (3): 193–206. https://doi.org/10.1080/15405702.2021.1913493.

O'Connor, Gabby. 2011. *What Lies Beneath, exh. cat.* Wellington, N.Z: City Gallery.

O'Connor, Gabby, and Craig Stevens. 2015. Combining Art and Science in a Primary School Setting: Paper and Ice. *Journal of Science Communication* 14 (04): A04:1–15. http://jcom.sissa.it/archive/14/04/JCOM_1404_2015_A04

———. 2018. STUDIOANTARCTICA: Embedding Art in a Geophysics Sea Ice Expedition. *Leonardo* 51: 57–58. https://doi.org/10.1162/LEON_a_01556.
Pearson, Mike. 2004. "No Joke in Petticoats": British Polar Expeditions and Their Theatrical Presentations. *TDR/The Drama Review* 48 (1): 44–59. https://doi.org/10.1162/105420404772990664.
Rauchfleisch, Adrian, Dario Siegen, and Daniel Vogler. 2021. How COVID-19 Displaced Climate Change: Mediated Climate Change Activism and Issue Attention in the Swiss Media and Online Sphere. *Environmental Communication*: 1–9. https://doi.org/10.1080/17524032.2021.1990978.
Schaffner, Franklin J. 1968. *The Planet of the Apes*, Arthur P. Jacobs, 112 minutes.
Selvey, Linda A., Morris Carpenter, Mattea Lazarou, and Katherine Cullerton. 2022. Communicating about Energy Policy in a Resource-Rich Jurisdiction during the Climate Crisis: Lessons from the People of Brisbane, Queensland, Australia. *International Journal of Environmental Research and Public Health* 19 (8): 4635. https://doi.org/10.3390/ijerph19084635.
Stevens, Craig, and Gabby O'Connor. 2016. Teleconnections in STEAM: Antarctic Field-Camp Art. *The STEAM Journal* 2 (2): 20. https://doi.org/10.5642/steam.20160202.20.
Stevens, Craig, Gabby O'Connor, and Natalie Robinson. 2019. The Connections Between Art and Science in Antarctica: Activating ScixArt. *Polar Record*. https://doi.org/10.1017/S0032247419000093.
Villeneuve, Denis. 2002. *Dune*, Warner Brothers, 156 mins.

Here Be Science Show Dragons: Ice, Icons and Metaphoric Approaches to Climate Change Communication

Graham J. Walker

Introduction

Ice and its antagonist fire are intimately linked with climate change, both through scientific phenomena and iconic or metaphoric representations. With newsfeeds awash with shrinking glaciers, burning fossil fuels, collapsing ice sheets and increasing bushfires the threat is all too clear, yet humanity is far from an effective, collective response. Communication of climate science alone has not been enough. Even if one takes climate change as fact, there is little correlation between awareness of the science and taking action—a recurrent pattern evident from early climate campaigns through to the present (Abrahamse and Matthies 2018; Staats et al. 1996; Whitmarsh et al. 2021). Significantly, this holds true for youth (Hanushek

G. J. Walker (✉)
Australian National University, Canberra, ACT, Australia
e-mail: graham.walker@anu.edu.au

A. Hemkendreis, A.-S. Jürgens (eds.), *Communicating Ice through Popular Art and Aesthetics*, Palgrave Studies in Media and Environmental Communication,
https://doi.org/10.1007/978-3-031-39787-5_4

and Woessmann 2012; OECD 2022)—the generation that will need to mop up rising emissions and melting ice. In response to this knowledge-behaviour disconnect, science education has suggested "interdisciplinary, affect-driven and experiential approaches" (Rousell and Cutter-Mackenzie-Knowles 2020, 196) while science communication has acknowledged the key role of affect, emotions and empowering narratives (Moser 2017; O'Neill and Whitmarsh 2009). Novel, experimental, motivational and transformative approaches to climate change communication are clearly needed. This chapter responds to this challenge with a focus on science shows.

Science Shows: Affect, Inspiration and Icy Origins

Science shows—a form of informal science learning popular in science centres and community/school outreach programmes—combine live demonstrations with engaging, enthusiastic delivery. A well-crafted show allows audiences to witness scientific phenomena firsthand, evoking a bouquet of emotions, as the performer(s) guides audiences through a narrative structured around scientific concepts, popular culture or—less often—socioscientific issues like climate change. Science shows differ from science theatre or plays by using live demonstrations as a central component, placing greater emphasis on audience interactivity and improvised content, and typically not using refined characters and elaborate dramatic techniques. With respect to popular culture, contemporary science shows are closer to busking performances, stand-up comedy, clowning and improvised theatre than tightly scripted plays or lectures. Their origin is, however, within the institution of science itself.

The forerunners of the modern science show were the lecture-cum-performances of Humphrey Davy and his successor, Michael Faraday, at London's Royal Institution during the nineteenth century (Sadler 2004; Walker 2012). Credit, however, is also due to performers who took science to the masses with a focus on entertainment, magic, wonder and illusion, such as the fascinating story of Dutchman L. K. Maju (da Rocha Gonçalves 2020). Regardless of setting, audiences were often less engaged with science's serious implications and more engaged by affect—quite directly in this case from the Royal Institution: "[Davy's] recommendation that nitrous oxide (laughing gas) be employed as an anaesthetic in minor surgical operations was ignored, but breathing it became the highlight of contemporary social gatherings" (Science History Institute 2017).

Like modern science shows, Davy provided his audience with firsthand, affective experience of phenomena. This included features key to contemporary shows, such as interactivity and audience volunteers—albeit with a more risqué approach to risk assessment. “There was Respiration, Nitrous Oxide, and unbounded Applause. Amen!” (quoted in Holmes 2008, 287), Davy penned on his early lectures in response to a volunteer breathing nitrous oxide—an unlikely demonstration in contemporary shows (but likely to get a laugh if it were). With Davy’s passing, Faraday—another brilliant scientist in his own right—continued to refine the format. He was an early champion of what we today call science communication—and its links to popular culture—with historians describing his goal as making “science a polite entertainment requiring the sort of suspension of disbelief that is associated with the theatre” (James 2002, 227). Faraday was an ardent proponent of communicating science through live demonstrations, declaring “If I said to my audience, ‘This stone will fall to the ground if I open my hand,’ I should open my hand and let it fall. Take nothing for granted as known; inform the eye at the same time as you address the ear” (Thompson 2005, 232).

Ice also featured at London’s Royal Institution, with shows providing early public engagement with cryogenics—the science of the extremely cold. In the late 1800s James Dewar experimented with the formation of ice, showing pressure lowered the freezing point of water (Dewar 1880). This was dramatic science: the point at which ice turned water and vice-versa had long been seen as a constant, and still is, which I argue is one reason this transformation holds such significance be it in science, art or beyond. It signifies change in systems viewed as stable. Dewar also performed demonstrations with incredibly cold liquid air (primarily liquid nitrogen, a common science show material today) which could instantaneously freeze water—a marvel in an age where naturally occurring ice would be carefully stored or even imported. Pouring liquid air into a beaker of water created frantic bubbling and clouds of fog and—as expanded upon later—people found their own meaning, relevance and inspiration:

> Agnes Marshall was more excited by what was going on in the beaker. As the air boiled off, the water left in the beaker turned to ice. Mrs Marshall was the most famous cook of her day, renowned especially for her ices and sorbets. Dewar had just shown her a way to make instant ice cream. (Pain, 2001)

While Dewar remains famous for his invention of the Dewar flask, the vacuum-flask invention featured in thermoses, Marshall innovated with ice cream machines, the first ice cream cone and even the 'ice cave' using ice and salt around a vessel to create an early refrigerator. Ice was the muse, with science as a partner, but the interpretation came down to what was meaningful for the individual and wider society. The role of ice in science communication and art today is not dissimilar.

The legacy of these early 'shows' at Royal Institution continues today through their Christmas Lectures, with Frank James noting their goal "has been and remains to inspire audiences, not necessarily to educate them" (James 2002, 227). With respect to climate change and the knowledge-behaviour disconnect discussed earlier, inspiring audiences may well be more important than educating them.

Modern science shows also typically prioritise inspiration and motivation over education. Fundamental to motivation is emotional engagement (Deckers 2010) and, as it was historically, emotions/states including enjoyment, curiosity, interest and surprise, and associated performer factors such as enthusiasm, humour and interactivity are associated with motivation across a broad range of modern science shows (Walker 2012; Walker et al. 2013). The nature of motivation in today's shows, however, is usually vague as opposed to targeted. That is, shows aim to positively influence attitudes towards science or broadly inspire careers and further study, rather than explicitly change behaviour or deal with environmental, health or social issues (Walker 2012). Nevertheless, science shows can create startling impact and behaviour change, even for complex socioscientific issues such as HIV/AIDS (Walker et al. 2013). Studies demonstrate science theatre, presentations and museum exhibitions can influence behavioural intentions concerning the environment (Sutter 2008), illicit drug use (Cartmill and Day 1997), smoking (Koster and Baumann 2005) and other health-related behaviours (Carney et al. 2009). Ice also plays a role in science shows aiming to broadly inspire, but is yet to realise its potential for more targeted motivational interventions, as we explore below.

Ice in Contemporary Science Shows

Modern science shows and other informal learning methods feature ice in many ways, from challenging students to pick up ice blocks with cotton using salt rather than a knot (give it a try if curious) to exploring the formation of ice at zero degrees. Liquid nitrogen, as first made in quantity by

Dewar, is a mainstay of science shows—at minus 196 degrees it can 'freeze' almost anything, while also boiling violently at room temperature, creating pressure to explode balloons, launch fog-erupting rockets or cause explosions (my personal favourite being the Teddy Bear Volcano). Ice augments many liquid nitrogen demonstrations (Fig. 1); some more novel examples involve creating an ice drinking sphere with a partially frozen water balloon or freezing my own thumb and smashing its fragile, icy innards with a hammer. Worry not, I have successfully performed it more than twice—but a good showman always saves some secrets for those at the show!

While these icy demonstrations are undoubtedly entertaining, research has also investigated how they educate, inspire and create meaning. Studies of shows in a 'subzero-temperature science centre' with primary school children demonstrate cognitive learning on changes of state (e.g. water to ice) and related concepts, transient effects on enjoyment of science and science career aspirations; however, no effect on the social implications of science (Caleon and Subramaniam 2005, 2007)—what I broadly term here as relevance. Other studies highlight the lay public's lack of everyday, lived connection to materials like liquid nitrogen; this can be addressed by contextualising it through the more relatable contexts of atmospheric gases and freezing properties (Tuah et al. 2010). In my own shows, I use

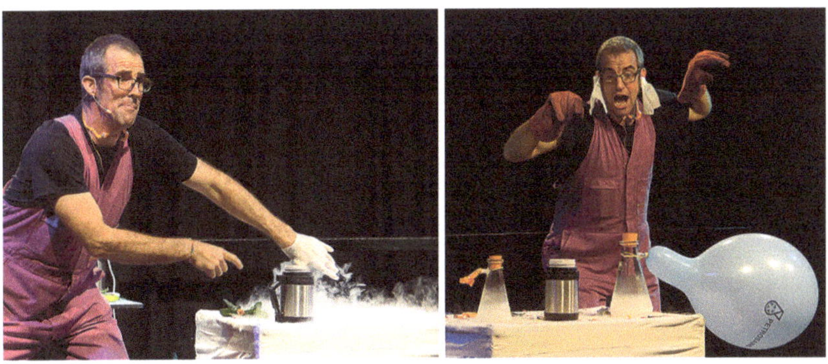

Fig. 1 Icy demonstration within a modern science show by Graham J. Walker, 2016 © private source (classic ice-themed liquid nitrogen science show demonstrations: freezing my thumb in liquid nitrogen revealing the main ingredient in humans is water (left) and using boiling liquid nitrogen to explode balloons—note the frosting of ice on the flasks (right))

the everyday ice-water-steam transition to explain changes of state in less common substances. In this way, ice becomes a bridge to broader understanding and insight in the entanglements between natural processes and everyday human lives, which brings us to climate change.

Using Science Shows to Communicate Climate Change

In 2001 I discovered science shows and science communication and was immediately hooked by the format of live experiments, bespoke homemade equipment, occasional explosions and sharing it all with an appreciative crowd. To this day I still tell people I have a career doing what I used to get in trouble for as a kid. After a decade seeing the effects shows have on audiences, hearing anecdotes of longer-term impact and pondering the potential of shows to address health and environmental problems, I undertook doctoral studies to better understand how shows can be tools for inspiration (Walker 2012). My central question was: Were these shows motivational and, if so, how, and in what domains?

I wrote and performed my first climate change show, *Sustain-ability! The Climate Change Show*, around 2010. The show explored the drivers of climate change, such as fossil fuels; impacts, like glacial and sea ice melting; renewable energies, such as solar and wind; and emerging technologies, like hydrogen (Fig. 2). The show applied lessons from behaviour change theory, such as leveraging social norms and targeting specific and realistic behaviour change (Ajzen 1991; Bamberg and Moser 2007). I aimed to craft an empowering, positive narrative that acknowledged the seriousness of climate change, but balanced that with what people could collectively and individually do to help. To inspire and engage, I avoided fear-based messaging (O'Neill and Nicholson-Cole 2009) and tried to counteract the negative feelings people typically associate with climate change (Moser 2014). This balance of climate risk/impact negatives—which are needed to prompt and design responses—and adaptation/mitigation positives—which are needed to inspire action and show positive possible futures—is now a central idea in climate communication. One must take care "not to communicate risks without also communicating possible solutions and accompanying cues that increase the listeners' sense of personal, group and response efficacy" (Moser 2017).

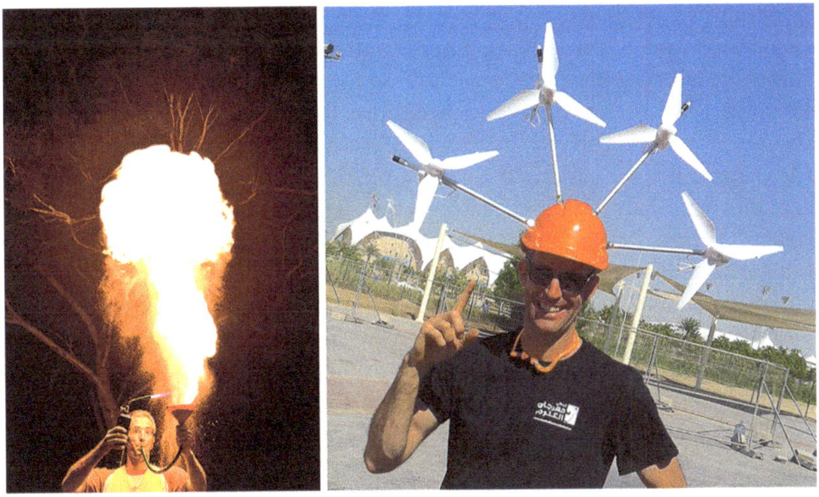

Fig. 2 *Sustain-ability! The Climate Change Show*. Science show dealing with climate change. Made by Graham J. Walker, 2010 © private source (the show presented both causes and solutions to climate change—the narrative used demonstrations such as burning emission-heavy fossil fuels like coal to make energy (left) followed by introducing emission-free technologies that can replace them such as wind turbines (right))

Science shows and other performance-based narrative formats have unique opportunities here, as the problem-solution balance can be woven into the narrative. Communicating science through narratives aids motivation, persuasion, environmental behaviour change, relevance, nuanced thinking, comprehension, emotional engagement and can have the effect of 'transportation'—being absorbed into a story and its persuasive power to influence beliefs (Avraamidou and Osborne 2009; Dahlstrom 2014; Rickard et al. 2021). Hence, narratives have the power to transform meaning and reveal novel futures—a tremendous tool in the communication of, and response to, climate change.

Moreover, well-crafted live performances are both captivating and, in a way, hold audiences captive—they are typically watched to their conclusion. This contrasts with climate communication via a website, online video or book where people often skim content or may not read/watch to the end. Recent meta-analyses on the effects of science communication narratives agree on their persuasive power but disagree on whether presentation medium (typically text, audio or video) affects persuasion (Braddock and Dillard 2016; Shen et al. 2015). They largely overlook, however, how people engage with these formats in the real world—one is far more likely to put down a brochure than walk out of a live show, though one can return to the brochure anytime. Only one of 99 studies in these meta-analyses focused on theatre; further research should unpack the motivational effects of captive but fleeting mediums, such as live shows. A show format—more than most other formats—offers opportunities for communicators to tell a story in full.

Pre-post and post-only surveys from *Sustain-ability! The Climate Change Show* showed people's motivation and specific environmental behavioural intentions were significantly improved, with audiences indicating they were more likely to recycle, compost, use renewable energy and use ethanol blended petrol after the show (Walker 2012). They also intended to exert more effort helping the environment and talk to friends about actions that can help, but fell short of intending to encourage others to adopt pro-environmental behaviours. Regression analysis showed motivation decreased with age; this may be an artefact of older audience members being more realistic and less subject to acquiescence bias, and/or, more optimistically, that younger generations are more willing to change. Intriguingly, statistical analysis showed that provoking emotions was not associated with increased motivation to act—the role of emotions was (statistically) usurped by judgements of the show's relevance, which was the most critical motivating factor.

Multiple studies (see Walker 2012) across multiple different shows, performers, topics and intended outcomes show relevance is by far the most critical aspect of science shows that aim to motivate. Comparable contexts, like science education, echo the potency of relevance on motivation, and highlight its cultural identity dimension (Hulleman and Harackiewicz 2009; Priniski et al. 2018). This key role of relevance and culture needs to be considered when using ice in science communication—a topic we will return to.

Beyond the data, *Sustain-ability! The Climate Change Show* gave a deeper lesson about branding and framing, relevance judgments, and about how people engage, or choose not to engage, when the words 'climate change' are highlighted in show names—marquees matter! Naming the show *Sustain-ability! The Climate Change Show* resulted in 15–25 people attending, whereas changing it to *Explosive Green Power* resulted in two full-houses of 180 people, with all other variables roughly the same.

It was a critical lesson: most people opting into free-choice (recreational) learning experiences do not want to be confronted by climate change—a topic most associated with doom and gloom, turning relevance into a negative. Rather, people want different framing of the issue, or completely new ways to relate to it—we need imaginative engagement that disrupts the status quo and repositions the relevance of climate issues. Creative and artistic approaches—particularly using iconic representations of climate change, like ice, as shared throughout this book—are hence a prudent response.

Making Meaning and Revealing Relevance: Framing, Icons and Cultural Nuance

Although I had worked hard to avoid making a 'preachy' show, I was forced to climb down off my climate change soapbox (or at least substitute it for a more cunning soapbox)—framing theory is a useful tool here. Put simply, framing is about choosing a context or focus in which to situate an issue—in other words, a relevance intervention. In science-art fusions, ice is commonly used as a frame to represent climate change (Gabrys and Yusoff 2012). Framing aims to set an agenda and influence how communication is interpreted and the meanings people construct from it. According to Nisbet and Mooney:

> Frames organize central ideas, defining a controversy to resonate with core values and assumptions. Frames pare down complex issues by giving some aspects greater emphasis. They allow citizens to rapidly identify why an issue matters, who might be responsible, and what should be done. (Nisbet and Mooney 2007, 56)

While effective for bums on seats, reframing responses to climate change as 'explosive green science' was a stretch for me; however, it did make me re-examine my show, its central narrative (a key framing device) and the meaning I hoped people would leave with. I subsequently changed the show's name to *Future Energy* and reworked the content so it was framed around energy production, a topic I judged to be less threatening, more relevant to day-to-day life and open to inspiring positive narratives.

Ongoing performances since this shift demonstrate the framing affects bookings, attendance, initial response and the meaning people construct—particularly evident in post-show conversations—and was relevant across cultures. This was a narrative about climate and what humans must change but put in less foreboding terms—it was about ubiquitous energy, different ways to make it, different resulting emissions and how we can save the planet at the same time. It was future-oriented and foregrounded hope, optimism and the temporal journey we are already on—and how we might collectively give it a happy ending. It proved Nisbet's contention that "communication is not simply a translation of facts but more importantly a *negotiation of meaning*" (Nisbet 2009, 41; italics added).

While artists have long used visual symbolism as a key negotiator of meaning that provides diverse and flexible opportunities for personal relevance, science communication has only recently explored such ideas—research on climate icons is particularly enlightening. The central theme of this book—ice—is a widely used climate change icon, with people signifying, summarising or directly measuring the impacts of climate change through images, stories or data regarding ice—such as retreating glaciers, collapsing ice sheets or the last refuge of stranded polar bears. In the context of the art-science intersections, Gabrys and Yusoff note "sites of ice loss are primary indicators of climate change that are materially evident, situated and local, while also inextricably linked to the seemingly abstract registers of global climate change" (2012, 6). Ice's iconic representation is a bridge for engagement. According to O'Neill and Hulme:

> An 'icon' [refers] to a tangible entity considered worthy of respect; something to which the viewer can relate and for which they feel empathy [...] a climate icon as defined here is a symbolic representation of more than what is immediately apparent. Thus, a climate icon represents more than simply an image, narrative or probability describing the entity which is being represented – an icon is the entity itself, bound up with how the viewer relates to that entity through their individual cultural values, world view and sense of place. (2009, 403)

This individual 'lens' that people see through when finding meaning in an icon suggests ice may not be the most relevant for specific communities; however, it remains a global climate change icon. O'Neill and Hulme's research in the United Kingdom used a robust participatory process to shortlist 141 climate icons. These were refined into three expert icons using scientific framing (e.g. changes in the West Antarctic Ice Sheet) and three non-expert icons using everyday framing (e.g. a polar bear—a widely used adjunct in ice iconography). Participants thought non-expert icons were more interesting and understandable than expert icons, and responses revealed three key themes as to why icons are personally meaningful: pragmatic/logical justifications, whether the spatial scale was local or distant, and intangible reasoning related to emotions or spirituality. While the second theme speaks to relevance as discussed earlier, the third theme conveys the complexity of how people find meaning in icons and, I argue, underscores the value of artistic approaches that target affect rather than logic (as described throughout this book).

Icons presented as a text and visual 'one-pager' increased people's perceptions of climate change as a serious issue/threat and their motivation to seek out climate information—in sum, icons are effective tools. Relevance due to geographic proximity and personal affinity was a strong influence (e.g. the icon of Norfolk Broads local nature area), however it was overruled by affect—polar-bears-on-ice image icons were judged overwhelmingly the least relevant, yet people were most drawn to them, "because they represented 'the idea of pure environment and fragile environment most affected by change'" (O'Neill and Hulme 2009, 407). This suggests ice-related icons resonate strongly with people, even if not directly relevant to them—but may be more potent if they are.

What an icon symbolises and the meaning people infer from it is highly dependent on whose eyes it is seen through—particularly their cultural context (Hodge and Kress 1988). Broader advice on climate communication stresses the power of local examples and lived experience (Moser 2017) and one would expect icons to operate in the same way. Key considerations of spatial/geographic relevance and affect provide guides as to how to target icons. Ice may resonate strongly with New Zealanders as glaciers retreat, however further north in Tuvalu, the iconography of rising sea level trumps all, with Pacific Island Forum delegates welcomed by children sitting in a moat of water (Lyons 2019). Scientifically, however, they are two sides of the same coin. In Samoa, mangroves take on iconic status—scientifically they protect from sea erosion and extreme weather, reduce nutrient runoff protecting corals, provide habitat and supply community resources (Suluvale 2001). Their personal meaning is arguably more powerful: mangroves are places where individuals played as children, saw their climate-resilient properties firsthand, found traditional medicine and materials and gave homes to tasty mud crabs reserved for chiefs and dignitaries (S. Ale, personal communication, 8 April 2022). Fire—a common phenomenon in my climate demonstrations—is a multifaceted climate icon, signifying fossil fuel consumption, resulting in emissions and climate impacts like bushfires. Fire, perhaps more than ice, is a confronting but relevant icon in my home of Australia; it sums up complex relationships between humans, energy and climate change (Clark and Yusoff 2014).

Tailoring icons to the audience's demographic characteristics is wise, but also provides a muse for creative and experimental approaches. Despite the success of my *Future Energy* show, it had limitations, particularly with mixed age family audiences. Younger children struggled to grasp the energy production narrative and underlying science—it was too expansive, remote and abstract, and was not intrinsically interesting or relevant to them. Children, and subsequently their parents, would sometimes disengage; they needed a different way to think about the issue.

Hence the show narrative was completely redesigned, this time employing metaphor as a framing device in consideration of the young audience. Firstly, the remoteness of the global energy transition was represented by a robot using different forms of energy for a 'Robots Got Talent' dance routine. Secondly, carbon dioxide emissions were represented by flatulent emissions from the robot character, CO_2PO (played by clown and character actor, Mark Johnson). Emissions are a central idea in climate science; symbolising them through a robot's dance-routine-wrecking flatulence produced a child-friendly iconic representation. The science and narrative around different sources of energy remained essentially the same, however the live demonstrations were integrated with the robot character (Fig. 3).[1]

The outcome—*Dr Watts' Robot Energy Show*—was well received by families; kids laughed at the 'fart jokes' and the robot-needs-energy narrative was relevant (prompting comparisons to a toy with flat batteries), while conversations suggested adults left with the energy transition message and valued the metaphor's kid-friendly framing. Nevertheless, in respect to O'Neill and Hulme's (2009) three themes of engaging climate icons, a farting robot in a dancing competition may have limited appeal beyond families with young children. While the robot-needs-energy narrative was logical/pragmatic, there was little direct spatial/geographic relevance (though energy issues related to the Australian audience's context), and the emotional dimension based on flatulent mirth and sympathy for the robot and its creator may be less effective, or even offensive, for adults or in different cultural settings. Further research should explore if these three themes apply in creative and artistic contexts, and whether affect can trump other aspects as it does in O'Neill and Hulme's (2009) study. This

[1] A trailer for the show can be viewed at https://www.youtube.com/watch?v=6sItrwBZzmk

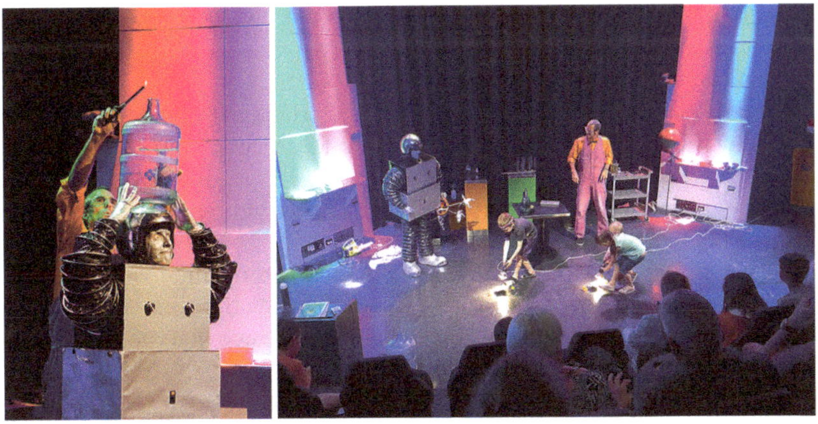

Fig. 3 *Dr Watts' Robot Energy Show*. Science show with focus on climate and energy transitions by Graham J. Walker, 2014 © private source (In *Dr Watts' Robot Energy Show* different sources of energy for the dancing robot CO_2PO (played by Mark Johnson) were used as a metaphorical representation of how energy is made for wider society, while CO_2 emissions were represented as the robot's flatulence—an issue that needed to be remedied for a winning dance routine in 'Robots Got Talent'. A biofuel energy source about to be ignited (left) was discussed as a transition fuel, while technologies like solar were first tested by members of the audience, providing interactivity (right), before being fitted to the robot.)

is a critical question for creatives and artists whose tools are primarily affective, but like science communicators strive for genuine engagement.

Ice and Other Climate Change Metaphors

Science shows do not have a tradition of metaphorical narrative, but I argue with climate change it may be helpful. People have long turned to metaphors to aid communication, understanding, memory and behaviour change, particularly when topics are complex, multifaceted or potentially polarising. This is certainly the case in climate change communication. Metaphor gives different ways to view climate change which may unlock new thinking, build empathy, reduce climate anxiety and in turn foster feelings of agency. While metaphor, particularly when used as an overarching narrative device, comes with risk—people may not connect the metaphor to specific actions they could take, or fail to see the underlying meaning completely—it is a strategy worth experimenting with given direct communication often falls flat.

Metaphor (as used here) is defined "as understanding one conceptual domain [the target] in terms of another conceptual domain [the source]" (Kövecses 2002, 4). This cognitive linguistic perspective views metaphor as more than a feature of language, but also part of how people think and behave (Lakoff and Johnson 1980). Building on the work of Black—who sees metaphor as a 'filter' to forge new understanding and meaning of the target domain, which is a 'system' implying relationships rather than things (Black 1977)—there is now extensive evidence that metaphor is an intrinsic part of cognition (Johnson 2010).

Arnold Modell (2009) goes further, arguing metaphor is the manifestation of neurophysiological processes that are reflected in language—this positions metaphor as a fundamental way we think. According to Modell, metaphor taps into the unconscious and its processes, unpacks emotional experiences and fosters empathy. Aligning how we communicate climate change with deeper processes people use to think about it appears a fruitful opportunity. Modell extends the definition given above to say metaphor "not only *transfers* meaning between different domains, but by means of novel recombinations metaphor can *transform* meaning" (Modell 2009, 6). Modell is advocating for greater emphasis on metaphor as a tool in psychoanalysis, particularly to address trauma and give patients increased agency—an aspiration equally applicable to climate change. Metaphor offers a way for climate change shows and other artistic approaches to align with how people approach these issues beyond rationality, tapping into emotion and empathy, and sidelining cognitive processes which often fail to bring about change.

Turning to related fields, visual, linguistic and hybrid metaphors have been employed in climate education (Niebert and Gropengiesser 2013), art-based climate activism (Cozen 2013), public engagement (O'Neill and Hulme 2009) and journalism and media (Atanasova and Koteyko 2017). Climate change metaphors are often deployed to shift people's thinking and actions, be it large-scale politics and diplomacy (Bernstein and Hoffmann 2019) or individual behaviour change (Flusberg et al. 2017). Indeed, metaphors' power in climate communication—particularly when fused with art—is that it "can rhetorically mediate new ways of being in the world" (Cozen 2013, 301).

This quote is telling and points to individuals' roles and different possible futures and—if art or science communication aims to inspire positive change—care must be taken when selecting future-oriented metaphors and framing. Take ice—from melting glaciers to reflecting solar radiation—which plays a key role in climate change. When brought together with art and (often) metaphor, ice forms a lens to understand these effects at the human scale. Artistic interpretations, however, tend to focus on impacts and negatives, but also show adaptation and mitigation possibilities (Gabrys and Yusoff 2012).

Given the psychological toll of climate change (American Psychological Association 2011) and literature reviewed earlier on impact-solution balance, I argue here for more emphasis on sharing positive futures through artistic and creative approaches—particularly metaphor. On a neuropsychological level, metaphor's strength is its ability to simplify complex and confronting issues like climate change, which reduces anxiety and promotes a more constructive relationship with the issue—this opens the mind to action and behaviour change (P. Sundaralingam, personal communication, 5 October 2022). This opportunity may be stifled if metaphors promote anxiety, a point creatives should bear in mind. While works like Olarfur Eliasson's *Ice Watch London* may be effective political protest and certainly raise awareness, prompting reviews like "Scary, isn't it?" (Delicata 2018), whether they help individuals cope, change and respond effectively is unclear.

In Conclusion

Ice has long been a source of fascination for humans—its melting, freezing and other physical properties demonstrate fundamental properties of matter, but its everyday relevance for people and significance as an indicator of both environmental consistency and change give it far deeper meaning. While one may seek to describe ice purely scientifically, people will often derive deeper meaning. In science shows, ice (and materials like it such as fire) features in this literal sense to communicate basic science, but also in contexts such as climate change. With that, ice, fire and other icons become more than facts and phenomena, but also prompt relevance and become frames and metaphors with potential to create new meaning and influence attitudes and behaviour. In artistic settings this has long been fluid, but in science shows these ideas are only beginning to thaw. Creative science communicators and science show performers need to experiment with metaphor and its potential to engage people more effectively with climate change. So, responding to calls by Gabrys, Yusoff and others, what might these imaginative art-science framings or new perspectives look like? I leave you with a show concept currently under (early) development, *Net-Zero Dragon*.

Dragons are notoriously difficult pets. Many owners complain about their voracious appetite for high-energy foods, such as coal and natural gas, and the way they express their appreciation after mealtime, with scorching fire breathing obliterating all in its path. Worse still, when served a delicious sorbet shaved directly from the coldest Antarctic ice sheet to soothe the heat, that too is consumed in the fiery emissions. Then there is the invisible and destructive carbon-based intestinal wind, slowly trapping heat and causing yet more problems, especially as dragons age—unless there is early dietary intervention. Can our hapless protagonist get his dragon onto a clean, green diet of biofuels in time? Or take a step that some may say is radical but others say is already doable: go 'dragon vegan' and shift to a fire-free diet based on only solar and wind? Apparently, this can drastically reduce emissions, turning fiery breath into cooling foggy wafts (simulated with liquid nitrogen). Or will this dragon still crave fire, opting instead for delicious and highly flammable hydrogen—all the fire but without the unsavoury emissions from the tail area? Or better still, have a fuel cell implant directly converting hydrogen into electrical sustenance—producing only clean energy and water, providing feedstock to restore the ice. Will the dragon adapt to the new diet, giving it a cooling breath and allowing melted ice and puddled rising water to recover, or will it simply continue on as it is? Join us on a mission to make a lovable mechanical pet dragon a sustainable member of the family; one human's quest—but a question for all humanity.

Bibliography

Abrahamse, Wokje, and Ellen Matthies. 2018. Informational Strategies to Promote Pro-Environmental Behaviour. In *Environmental Psychology: An Introduction*, ed. Linda Steg and Judith de Groot, 261–272. Hoboken: John Wiley & Sons, Ltd. https://doi.org/10.1002/9781119241072.CH26.

Ajzen, Icek. 1991. The Theory of Planned Behavior. *Organizational Behavior and Human Decision Processes* 50 (2): 179–211. https://doi.org/10.1016/0749-5978(91)90020-T.

American Psychological Association. (2011). Psychology & Global Climate Change: addressing a multifaceted phenomenon and set of challenges. https://doi.org/10.1039/c1em10131e

Atanasova, Dimitrinka, and Nelya Koteyko. 2017. Metaphors in Guardian Online and Mail Online Opinion-Page Content on Climate Change: War, Religion, and Politics. *Environmental Communication* 11 (4): 452–469. https://doi.org/10.1080/17524032.2015.1024705.

Avraamidou, Lucy, and Jonathan Osborne. 2009. The Role of Narrative in Communicating Science. *International Journal of Science Education* 31 (12): 1683–1707. https://doi.org/10.1080/09500690802380695.

Bamberg, Sebastian, and Guido Moser. 2007. Twenty Years After Hines, Hungerford, and Tomera: A New Meta-analysis of Psycho-social Determinants of Pro-environmental Behaviour. *Journal of Environmental Psychology* 27: 14–25. https://doi.org/10.1016/j.jenvp.2006.12.002.

Bernstein, Steven, and Matthew Hoffmann. 2019. Climate Politics, Metaphors and the Fractal Carbon Trap. *Nature Climate Change* 9 (12): 919–925. https://doi.org/10.1038/s41558-019-0618-2.

Black, Max. 1977. More About Metaphor. *Dialectica* 31 (3): 431–457.

Braddock, Kurt, and James P. Dillard. 2016. Meta-analytic Evidence for the Persuasive Effect of Narratives on Beliefs, Attitudes, Intentions, and Behaviors. *Communication Monographs* 83 (4): 446–467. https://doi.org/10.1080/03637751.2015.1128555.

Caleon, Imelda, and Ramanathan Subramaniam. 2005. The Impact of a Cryogenics-Based Enrichment Programme on Attitude Towards Science and the Learning of Science Concepts. *International Journal of Science Education* 27 (6): 679–704. https://doi.org/10.1080/09500690500038306.

———. 2007. Augmenting Learning in an Out-of-School Context: The Cognitive and Affective Impact of Two Cryogenics-Based Enrichment Programmes on Upper Primary Students. *Research in Science Education* 37 (3): 333–351. https://doi.org/10.1007/S11165-006-9032-7.

Carney, Patricia A., Arwen Bunce, Nancy Perrin, Linda C. Howarth, Susan Griest, Phyllis Beemsterboer, and William E. Cameron. 2009. Educating the Public About Research Funded by the National Institutes of Health Using a Partnership Between an Academic Medical Center and Community-Based Science Museum. *Journal of Community Health* 34 (4): 246–254. https://doi.org/10.1007/s10900-009-9150-z.

Cartmill, Randy S., and Lucille L. Day. 1997. Can Museums Make a Difference? *Curator: The Museum Journal* 40 (3): 197–210. https://doi.org/10.1111/j.2151-6952.1997.tb01303.x.

Clark, Nigel, and Kathryn Yusoff. 2014. Combustion and Society: A Fire-Centred History of Energy Use. *Theory, Culture & Society* 31 (5): 203–226. https://doi.org/10.1177/0263276414536929.

Cozen, Brian. 2013. Mobilizing Artists: Green Patriot Posters, Visual Metaphors, and Climate Change Activism. *Environmental Communication* 7 (2): 297–314. https://doi.org/10.1080/17524032.2013.777353.

Da Rocha Gonçalves, Dulce. 2020. Science Between the Fairground and the Academy: The Case of Dutch science Popularizer L. K. Maju (1823–1886). *Public Understanding of Science* 29 (8): 096366252096509. https://doi.org/10.1177/0963662520965093.

Dahlstrom, Michael F. 2014. Using Narratives and Storytelling to Communicate Science with Nonexpert Audiences. *Proceedings of the National Academy of Sciences* 111 (Supplement 4): 13614–13620. https://doi.org/10.1073/pnas.1320645111.

Deckers, Lambert. 2010. *Motivation: Biological, Psychological and Environmental*. 3rd ed. Boston: Allyn & Bacon.

Delicata, Hope. 2018. Olafur Eliasson on His Daring Melting Ice Blocks Installation. *Dazed*. https://www.dazeddigital.com/art-photography/article/42569/1/olafur-eliasson-ice-watch-installation-tate-london-climate-change. Accessed 25 Nov 2022.

Dewar, James. 1880. On the Lowering of the Freezing-Point of Water by Pressure. *Proceedings of the Royal Society of London* 30: 533–538. https://doi.org/10.1098/rspl.1879.0156.

Flusberg, Stephen J., Teenie Matlock, and Paul Thibodeau. 2017. Metaphors for the War (or Race) Against Climate Change. *Environmental Communication* 11 (6): 769–783. https://doi.org/10.1080/17524032.2017.1289111.

Gabrys, Jennifer, and Kathryn Yusoff. 2012. Arts, Sciences and Climate Change: Practices and Politics at the Threshold. *Science as Culture* 21 (1): 1–24. https://doi.org/10.1080/09505431.2010.550139.

Hanushek, Eric A., and Ludger Woessmann. 2012. Do Better Schools Lead to More Growth? Cognitive Skills, Economic Outcomes, and Causation. *Journal of Economic Growth* 17 (4): 267–321. https://doi.org/10.1007/s10887-012-9081-x.

Hodge, Robert, and Gunther Kress. 1988. *Social Semiotics*. New York: Cornell University Press.

Holmes, Richard. 2008. *The Age of Wonder: How the Romantic Generation Discovered the Beauty and Terror of Science*. London: Harper Press.

Hulleman, Chris S., and Judith M. Harackiewicz. 2009. Promoting Interest and Performance in High School Science Classes. *Science* 326 (5958): 1410–1412. https://doi.org/10.1126/science.1177067.

James, Frank A.J.L. 2002. "Never Talk About Science, Show It to Them": The Lecture Theatre of the Royal Institution. *Interdisciplinary Science Reviews* 27 (3): 225–229. https://doi.org/10.1179/030801802225003178.

Johnson, Mark. 2010. Metaphor and Cognition. In *Handbook of Phenomenology and Cognitive Science*, ed. Daniel Schmicking and Shaun Gallagher, 401–414. Dordrecht: Springer.

Koster, Emlyn H., and Stephen H. Baumann. 2005. Liberty Science Centre in the United States: A Mission Focussed on External Relevance. In *Looking Reality in the Eye: Museums and Social Responsibility*, ed. Robert R. Janes and Gerald T. Conalty, 85–112. Calgary: University of Calgary Press.

Kövecses, Zoltan. 2002. *Metaphor*. New York: Oxford University Press.

Lakoff, George, and Mark Johnson. 1980. Conceptual Metaphor in Everyday Language. *Journal of Philosophy* 77 (8): 453–486.

Lyons, Kate. 2019. Pacific Islands Forum: Tuvalu Children Welcome Leaders with a Climate Plea. *The Guardian*. https://www.theguardian.com/world/2019/aug/11/pacific-islands-forum-tuvalu-children-welcome-leaders-with-a-climate-plea. Accessed 25 Nov 2022.

Modell, Arnold H. 2009. Metaphor – The Bridge Between Feelings and Knowledge. *Psychoanalytic Inquiry* 29 (1): 6–11. https://doi.org/10.1080/07351690802246890.

Moser, Susanne C. 2014. Communicating Adaptation to Climate Change: The Art and Science of Public Engagement When Climate Change Comes Home. *Wiley Interdisciplinary Reviews: Climate Change* 5 (3): 337–358. https://doi.org/10.1002/wcc.276.

Moser, Susanne C. 2017. Communicating Climate Change Adaptation and Resilience. In *Oxford Encyclopedia on Climate Change Communication*, ed. Matthew C. Nisbet. London: Oxford University Press

Niebert, Karl, and Harald Gropengiesser. 2013. Understanding and Communicating Climate Change in Metaphors. *Environmental Education Research* 19 (3): 282–302. https://doi.org/10.1080/13504622.2012.690855.

Nisbet, Matthew C. 2009. Framing Science: A New Paradigm in Public Engagement. In *Communicating Science: New Agendas in Communication*, ed. Lee Kahlor and Patricia Stout, 40–67. New York: Routledge.

Nisbet, Matthew C., and Chris Mooney. 2007. Framing Science. *Science* 316 (5821): 56. https://doi.org/10.1002/cncy.20155.

O'Neill, Saffron, and Mike Hulme. 2009. An Iconic Approach for Representing Climate Change. *Global Environmental Change* 19 (4): 402–410. https://doi.org/10.1016/j.gloenvcha.2009.07.004.

O'Neill, Saffron, and Sophie Nicholson-Cole. 2009. "Fear Won't Do It": Promoting Positive Engagement with Climate Change Through Visual and Iconic Representations. *Science Communication* 30 (3): 355–379. https://doi.org/10.1177/1075547008329201.

O'Neill, Saffron, and Lorraine Whitmarsh. 2009. Encouraging Climate Change Engagement Through Affective Connections. *IOP Conference Series: Earth*

and Environmental Science 6 (57): 572003. https://doi.org/10.1088/1755-1307/6/7/572003.
OECD. 2022. Teaching for Climate Action. *Teaching in Focus* 44: 1–6. https://doi.org/10.1787/d3a72e77-en.
Pain, S. 2001. Queen of Ices. *New Scientist*, August 2001. https://www.newscientist.com/article/mg17123045-300-queen-of-ices/. Accessed 25 Nov 2022.
Priniski, Stacy J., Cameron A. Hecht, and Judith M. Harackiewicz. 2018. Making Learning Personally Meaningful: A New Framework for Relevance Research. *The Journal of Experimental Education* 86 (1): 11–29. https://doi.org/10.1080/00220973.2017.1380589.
Rickard, Laura N., Janet Z. Yang, Sixiao Liu, and Tabitha Boze. 2021. Fish Tales: How Narrative Modality, Emotion, and Transportation Influence Support for Sustainable Aquaculture. *Science Communication* 107554702098755. https://doi.org/10.1177/1075547020987555.
Rousell, David, and Amy Cutter-Mackenzie-Knowles. 2020. A Systematic Review of Climate Change Education: Giving Children and Young People a 'Voice' and a 'Hand' in Redressing Climate Change. *Children's Geographies* 18 (2): 191–208. https://doi.org/10.1080/14733285.2019.1614532.
Sadler, Wendy J. 2004. Evaluating the Short and Long-Term Impact of an Interactive Science Show. *Research for All* 5 (2): 399–419. https://doi.org/10.14324/RFA.05.2.14.
Science History Institute. (2017). Humphry Davy. https://www.sciencehistory.org/historical-profile/humphry-davy. Accessed 25 Nov 2022.
Shen, Fuyuan, Vivian C. Sheer, and Ruobing Li. 2015. Impact of Narratives on Persuasion in Health Communication: A Meta-analysis. *Journal of Advertising* 44 (2): 105–113. https://doi.org/10.1080/00913367.2015.1018467.
Staats, H.J., A.P. Wit, and C.Y.H. Midden. 1996. Communicating the Greenhouse Effect to the Public: Evaluation of a Mass Media Campaign from a Social Dilemma Perspective. *Journal of Environmental Management* 46 (2): 189–203. https://doi.org/10.1006/jema.1996.0015.
Suluvale, Eletise A. 2001. Environmental Change of Selected Mangrove Areas in Samoa. In *Samoan Environment Forum*, ed. Ministry of Natural Resources & Environment, 14–23. Apia: SPREP.
Sutter, Glenn C. 2008. Promoting Sustainability: Audience and Curatorial Perspectives on the Human Factor. *Curator: The Museum Journal* 51 (2): 151–202. https://doi.org/10.1111/j.2151-6952.2008.tb00305.x.
Swim, Janet, Susan Clayton, Thomas Doherty, Robert Gifford, George Howard, Joseph Reser, Paul Stern, and Elke Weber. 2011. *Psychology & Global Climate Change: Addressing a Multifaceted Phenomenon and Set of Challenges.* Washington DC: American Psychological Association.
Thompson, Silvanus. 2005. *Michael Faraday His Life and Work.* London: Kessinger Publishing (Originally Published 1901).

Tuah, Jauyah, Timothy G. Harrison, and Dudley E. Shallcross. 2010. A Review of the Use of Demonstration Lectures in the Promotion of Positive Attitudes Towards, and the Learning of Science with Reference to 'A Pollutant's Tale', a Demonstration Lecture on Air Quality and Climate Change. *Romanian Journal of Education* 1 (3–4): 93–102.

Walker, Graham J. 2012. *Motivational Features of Science Shows.* Australian National University.

Walker, Graham J., Susan M. Stocklmayer, and Will J. Grant. 2013. Science Theatre: Changing South African Students' Intended Behaviour Toward HIV AIDS. *International Journal of Science Education, Part B* 3 (2): 101–120. https://doi.org/10.1080/09500693.2011.633939.

Whitmarsh, Lorraine, Wouter Poortinga, and Stuart Capstick. 2021. Behaviour Change to Address Climate Change. *Current Opinion in Psychology* 42: 76–81. https://doi.org/10.1016/j.copsyc.2021.04.002.

Ice Exploration: Heroism, Art and Imaginaries

Ethnography as Racialised Womanhood in the Arctic Writings of Josephine Diebitsch-Peary

Nanna Katrine Lüders Kaalund

Introduction

In 1892, an American newspaper congratulated Josephine Diebitsch-Peary (referred to as Diebitsch-Peary from hereon) on her safe return to the US "from such a trip as no woman every essayed before" (The Iola Register 1892). Diebitsch-Peary travelled to the Arctic with her husband Robert Peary (referred to as Robert from hereon) and a small team, and she spent the winter of 1891–1892 in McCormick Bay, Canada, while

The work presented here is supported by the Carlsberg Foundation, grant CF20-0180.

N. K. L. Kaalund (✉)
Department of Philosophy and History of Ideas, Aarhus University, Aarhus, Denmark
e-mail: nklk@cas.au.dk

A. Hemkendreis, A.-S. Jürgens (eds.), *Communicating Ice through Popular Art and Aesthetics*, Palgrave Studies in Media and Environmental Communication,
https://doi.org/10.1007/978-3-031-39787-5_5

Robert attempted to reach the North Pole. This was her first of several Arctic expeditions. She published a travel narrative, *My Arctic Journal* (1893) in addition to two children's books—*The Snow Baby* (1901) and *Children of the North* (1903)—based on her experiences.

As a woman writer and traveller, Diebitsch-Peary navigated boundaries of authority that were traditionally gendered as masculine, both in the context of Arctic exploration and in American lecture halls (Bergmann 1993; Bloom 1993; Erikson 2013; Hansson and Norberg 2009; Reeploeg 2019). Diebitsch-Peary's decision to visit the Arctic did indeed raise some eyebrows. "Great was the surprise when the public learned that the wife of Lieutenant Peary was to accompany her husband to North Greenland", wrote one American newspaper, noting that "such a journey was regarded as an anomaly in the annals of Arctic exploration" (The Morning Call 1891). By virtue of her gender, Diebitsch-Peary could not easily draw on the well-known tropes of the heroic male explorer when portraying herself as an explorer or as an expert on Arctic matters. Instead, Diebitsch-Peary invoked a different form of heroism, one that was centred on a performative domesticity as she created an American home in the Arctic. This was a home Diebitsch-Peary carefully contrasted to everything and everyone she encountered in the Arctic—a home that represented Westernised civility in a frozen landscape she described as inherently uncivilised.

To justify her presence in the Arctic, Diebitsch-Peary drew on the popular imperialist ideologies that positioned the white American woman as a feminine bastion of civilisation in frontier settlements—what scholar Amy Kaplan has termed "manifest domesticity" (Kaplan 1998). Drawing on recent literature on women's travel writing, the history of the human sciences and Arctic studies, this chapter shows how Diebitsch-Peary created a space for herself as an Arctic explorer and communicator by repurposing the masculine tropes of the heroic explorer in a racialised language of American domesticity.

Diebitsch-Peary had broad audiences and diverse venues of communication, ranging from lectures at scientific societies to children's books directed at families. Her publications and lectures were an important source of income and cultural capital for herself and her husband. For example, Diebitsch-Peary embarked on a lecture tour to raise funds for a search expedition when it appeared that one of Robert's later North Pole ventures had gone awry. Her lectures, newspaper interviews and publications gathered attention and raised goodwill towards Robert's further efforts to reach the North Pole. She had connections to the learned societies and institutions, and secured sponsorships and financial contributions from several of these groups, including the Brooklyn Institute of Arts and Sciences and the National Geographical Society.

At the same time, Diebitsch-Peary's path to being accepted as an expert on Arctic matters was not straightforward. At the turn of the twentieth century, Diebitsch-Peary was frequently described as "the Arctic explorer's wife" (Freeland Tribune 1884), as Robert's "trusty comrade and indefatigable assistant" (The Sunday Star 1909), or simply as "Mrs Robert E. Peary" (New-York Daily Tribune 1902). This bias is also reflected in the literature on Arctic exploration. As historian Silke Reeploeg convincingly shows, the memorialisation of Diebitsch-Peary has been shaped by the gendered practices of colonial archives and the broader power-structures of colonialism, which favoured specific voices—especially white males—over others (Reeploeg 2019, 2021). By taking seriously Diebitsch-Peary's broad authorship as a form of science communication, and as an early example of American women's field ethnography, this chapter reveals how the interconnection of gender, race and climate shaped the American efforts to claim the North Pole.

The Arctic home Diebitsch-Peary established was also an important observational site that formed the basis of the ethnographic data in her highly popular works. In order to examine how Diebitsch-Peary's ethnography was shaped by an ideology of manifest domesticity, it is necessary to first critically interrogate how her presence in the Arctic challenged well-established conceptions of Arctic exploration. Diebitsch-Peary was, of course, not the only Euro-American woman in the Arctic, and her experiences were not unique simply by virtue of her gender and race. In the first section, I show how it was through a curated combination of colonial identities—being a white, educated and relatively well-to-do woman—that her experiences were made into something that was attention worthy.

Diebitsch-Peary's efforts to mitigate the criticism of her presence in the Arctic show the difficulties of portraying Arctic travel as safe, while maintaining the heroism associated with Arctic exploration, which justified both her presence and Robert's attempts to reach the furthest north. In the second section, I focus on the home in Diebitsch-Peary's ethnographic practices. Whereas Robert's manifest destiny was reaching the North Pole, hers was 'keeping house' in a place she described as lonely, uncivilised and frozen. This, in turn, formed the backdrop for her ethnographic methodologies. In the final section, I show how Diebitsch-Peary's descriptions of ice and snow—coming together in a hostile and sublime beauty—emerge as the key defining features of Arctic life. Culture and civilisation, in Diebitsch-Peary's accounts, belonged to the Americans. So did the North Pole.

Gendering the Tropes of Arctic Exploration

There is no shortage of stories about the men who left warmer shores for the Arctic. These are tales that are based on first-hand experience, fantasy and a combination of both. In these accounts, the explorer is typically portrayed as a man who heroically faces the perils of a dangerous icy landscape, to discover new land and scientific knowledge, and bring glory to their homeland. It is historiographically well-established that such ideals about exploration were imperialistic constructs that do not reflect the genuine lived experiences of the so-called explorers (David 2000; Grace 1997; Hanrahan 2017; Heggie 2019; Hill 2009). Yet, these tropes are enduring, especially in the English-speaking context. It is therefore not surprising that Diebitsch-Peary's presence in the Arctic does not easily fit into well-established conceptions of the explorer or of the Arctic as a field-site for explorations and discovery (Bergmann 1993; Bloom 1993; Grace 1997). The way in which Diebitsch-Peary negotiated these tropes brings to the fore the labour that went into establishing and upholding desired visions of Arctic exploration. This is particularly poignant as Diebitsch-Peary travelled during a period that saw significant changes in the relationship between science and travel, combined with growing opportunities for women to participate in it (Alacovska 2015; Hansson and Norberg 2009; Reeploeg 2019).

Diebitsch-Peary was born in Maryland in 1863 and grew up in Washington DC. Her father, Herman Diebitsch, worked at the Smithsonian Institution, an important centre of scientific research and natural history collection. Through her parents, Diebitsch-Peary was introduced early to the vibrant research culture in Washington. She attended the Spencerian Business College and graduated valedictorian in 1880. After graduation, Diebitsch-Peary first worked at the U.S. Census Bureau and later took over her father's role as a clerk at the Smithsonian Institution. It was a well-connected Diebitsch-Peary who Robert met in 1882, and after they married, Diebitsch-Peary drew on her training, employment and connections to lobby for support for her and Robert's Arctic ventures.

Though women were still generally excluded from holding positions at universities and being elected fellows to scientific societies, there were many opportunities for women to participate in science more broadly (Larsen 2017; Lind 2020; Shteir 1997). Women science writers in the nineteenth century employed a plethora of genres and styles of writing to establish themselves as authorities, and to sell books. As Bernard Lightman has shown, one of the popular strategies used by women science writers was to make their work accessible to a broader reading audience (Lightman 2009). Making science accessible for her contemporary audiences included avoiding scientific terms and systems of scientific classifications, or choosing other literary formats aside from scientific journal articles. This was one of the communicative strategies employed in Diebitsch-Peary's in her *My Arctic Journal*, and even more so in her two children's books (Aarekol 2020).

As opportunities in both scientific research and authorship increased in the later decades of the nineteenth century, more women authors experimented with genres outside of those that had traditionally been gendered female, like the so-called maternal tradition of science writing or children's books (Browne 1994; Larsen 2017; Lightman 2009, 95–166; Shteir 1997, 236–255). As historian Michael Robinson has observed, there was a significant growth in women writers in the US during the 1870s and 1880s, and this increase shaped the literary marketplace (2015). Yet, Diebitsch-Peary chose to publish a travel narrative and children's books—genres that had long been used by women (Hansson et al. 2020). What made Diebitsch-Peary's publications stand out was the subjects, rather than the familial format. Historian Mary Louise Pratt notes that "while women writers were authorised to produce novels, their access to travel writing seems to have remained even more limited than their access to travel itself, at least when it came to leaving Europe" (Pratt 1992, 103). Yet, as Pratt's analysis further shows, 'imperial eyes' could be both male and female travellers, which afforded different perspectives on the land and peoples encountered.

In her travel narrative, the reader follows Diebitsch-Peary as she embarks on a journey that had previously been reserved for male travellers. In her children's books, the subject of Arctic exploration is even further subverted, as the main character was Diebitsch-Peary's daughter Marie Ahnighito Peary. Marie was born during Diebitsch-Peary's second visit to Greenland, and later returned to the Arctic with her mother to meet her father. Just as Diebitsch-Peary used literary formats that were traditionally gendered as female, she also repeatedly presented herself as a submissive woman, who only travelled to the Arctic to support her husband. She gave numerous newspaper interviews where she told reporters "how she provided for the comforts" of the expedition, specifically by "keeping house" (Pittsburgh Dispatch 1892). Diebitsch-Peary's choice of literary format and the way she described her role during these ventures should be seen both within the context of women's science writing and as part of the changing relationship between travel and scientific practice in the Arctic around the turn of the twentieth century. As Diebitsch-Peary and her husband sought to generate financial and cultural support for Robert's attempts to reach the North Pole, it was important to portray Robert and his ventures as achievable.

Robert was convinced that previous attempts to reach the North Pole had failed in part because the explorers had not adopted Inuit methods for travelling and living in the Arctic. The notion that Arctic travel could be safe, even relatively comfortable, was significant. It formed part of Robert's efforts to raise financial, cultural and scientific support. Because so many past expeditions had failed and/or experienced severe difficulties and sufferings on route to the icy north, the American government was no longer willing to financially support more of these ventures. It was important for Robert to show that his venture would be different: it was not going to be a disastrous waste of funds, time and life. His proposed methods for travelling were, he argued, both safer and more cost-efficient. From this perspective, Diebitsch-Peary's presence in the Arctic, especially when pregnant or with an accompanying child, supported the argument that Arctic travel was not as dangerous as previously assumed.

Despite arguing that Arctic travel was much safer than suggested by the experiences of previous expeditions, the heroic explorer trope was still significant in both Diebitsch-Peary's and Robert's Arctic work. There is a tension here. While both Diebitsch-Peary and Robert framed their Arctic travels as relatively safe, they still drew on the older rhetoric of Arctic exploration as heroic in the face of immense danger. Diebitsch-Peary's self-portrayal as a supportive wife and domestic anchor for Robert in the Arctic was key to balancing this tension between safety and heroism.

My Arctic Journal was published during the couple's second joint visit to the Arctic. There had been several critical articles published about their ventures in the general periodical press. These articles focused on the gendered dynamics of the couple. For example, in October 1893, a gossip column was reprinted in several newspapers with the headline: "Mrs. Peary is boss". The column reported on the content of a letter supposedly sent by Diebitsch-Peary to an anonymous friend in New York. Diebitsch-Peary was said to have defended "herself playfully from the charge of having reduced her husband to a henpecked condition" (The Indianapolis Journal 1893; The Morning News 1893; Wheeling Register 1893). It was well known, the anonymous author remarked, that Robert was "absolutely controlled by his wife" and this had "much impeded" his plans for a second expedition. Though Diebitsch-Peary was quoted as having refuted these allegations, the gossip column did not believe this statement to be true. "Bosh! Everybody who has had anything to do with this Peary expedition knows that the wife, and not the husband, was in command", exclaimed the anonymous author. The implication was that Robert was emasculated by the "arctic queen" (The Indianapolis Journal 1893; The Morning News 1893; Wheeling Register 1893).

Such gossip reveals the complexity of Diebitsch-Peary's challenges in establishing an identity as a woman Arctic explorer. It also shows that it was a challenge for Robert to navigate what travelling with his wife said about him and his proposed ventures. Was Robert actually going to do everything he could to reach the furthest north, or was he too "henpecked" to leave his wife behind for extended periods? Was Diebitsch-Peary the real commander? In the context of this gendered criticism of both Diebitsch-Peary and Robert's Arctic identities, Diebitsch-Peary's repurposing of the manifest destiny ethos of American colonial expansionism into (Arctic) domesticity becomes clear. The focus in her interviews, lectures and publications was always on the domestic work she undertook, and this labour was—as the following section shows—contrasted to the domestic traditions of Inuit women. Diebitsch-Peary's descriptions of herself and her husband emphasised the differences between their roles during the expedition, as well as their different forms of heroism. Hers was a special type of womanly bravery, where she supported her husband at all costs, and anxiously waited for him to return from his heroic exploratory missions. Diebitsch-Peary's publications and lectures directed public opinion about herself and Robert, and the ethnographic research she undertook while in her Arctic home was the basis for this performance of gendered Arctic identities.

The Home as an Observatory

There is a long tradition of women undertaking ethnographic and anthropological work (Larson 2011; Sera-Shriar 2013; Vider 2021). Many women worked with a male partner, usually their husband. As Edward

B. Tylor stated at an address to the Anthropological Society of Washington in 1884, "the man of the house, though he can do a great deal, cannot do it all. If his wife sympathises with his work, and is able to do it, really half the work of investigation seems to me to fall to her, so much is to be learned through the women of the tribe, which the men will not readily disclose" (as cited in Visweswaran 1997, 597). As Tylor's comment indicates, women researchers, by virtue of their gender, were able to access cultural aspects of societies that their male counterparts were not privy to. This difference in cultural access is seen in Diebitsch-Peary work. For example, the publisher's note to *My Arctic Journal* remarked: "The opportunities which Mrs. Peary had of observing [Inuit] manners and mode of life have enabled her to make a valuable contribution to ethnological learning" (Diebitsch-Peary 1893, 2). The claim was that as a woman, Diebitsch-Peary had privileged access to intimate aspects of Inuit domestic culture, which enabled ethnographic observations that had not been documented by male explorers. In this way, Diebitsch-Peary could combine the tradition of ethnographic study with the distinctively masculine role of the Arctic explorer.

To Diebitsch-Peary and many other ethnologists in the late nineteenth century coming out of the American tradition, extra-European peoples were studied as part of the natural environment. Researchers such as Franz Boas believed that to understand the history of Arctic peoples was to understand early human migration patterns, as well as the development of specific physical traits and cultural and religious practices. This methodological framework is known as diffusionism. The debates surrounding human history in the second half of the nineteenth century involved categorising all of humanity into different developmental frameworks. The Indigenous peoples of northern Greenland, Inughuit, were of particular interest in these debates (Rud 2013, 2017). Northern Greenland was, and still is, often conceptualised by Europeans and euro-Americans as the edge of the world—the Ultima Thule.

My Arctic Journal did not contain theoretical details on ethnography, and Diebitsch-Peary did not explicitly situate her observations within the contemporaneous theories of human physical and cultural development. There was no reference to people such as Boas, Otis Mason and Samuel George Morton, nor to the numerous other so-called explorers who had encountered and described Arctic Indigenous peoples during their voyages. Readers of her children's books would not have expected such details either. Diebitsch-Peary followed the strategy of many women science writers in this period: avoiding technical terms or references to concurrent scientific debates. Readers had to infer Diebitsch-Peary's position on topics debated in the human race sciences (which, at the time, encompassed anthropology and ethnography). Although Diebitsch-Peary did not include traditional scientific terminologies in her work, her position on the intersection between culture, civilisation and the environment is clear. Unlike Boas (a figurehead for cultural relativism), Diebitsch-Peary's portrayal of Inuit was highly environmentally deterministic; they were seen as products of their environment and biologically limited by race, in the American tradition of Samuel Morton and Louis Agassiz (Menand 2001–2002; Barth et al. 2005).

Diebitsch-Peary's writings were explicitly billed as ethnographic with an emphasis on her skills as a field-based expert, even though she did not include theoretical details to substantiate her research. As Robert wrote in the preface to *My Arctic Journal*, Diebitsch-Peary had "been where no white woman has ever been [...] and she has seen phases of life of the most northerly tribe of human beings on the globe, and in many ways has been enabled to get a closer insight into their ways and customs than had been obtained before" (1893, 3–4). Contemporary reviews of *My Arctic Journal* demonstrate that Peary had, at least partially, succeeded in creating this identity for herself. One newspaper noted that "it is evident that she is a woman of uncommon physique, fortitude and pluck" and that "the very complete account of those amiable people will be of special interest to the ethnologist; perhaps also the entomologist" (The Times 1893). Another remarked that the narrative was characterised by a "simple directness" and "courage and endurance" (The Salt Lake Herald 1894). The female authorship of the book was repeatedly emphasised in reviews, highlighting that the book was "unique" as the "bulk of it is written by a lady" (The New-York Tribune 1893; The Scottish Geographical Magazine 1894).

In addition to the value human race science placed on observations of Inuit, learning about the local communities was important to explorers for more direct reasons. For the sake of current and future expeditions, many explorers—including Diebitsch-Peary and Robert (Harper 1986, 2002)—explicitly sought to establish (what they described as) 'friendly relations' with Indigenous peoples, exploiting Inuit labour and knowledge to help the expeditions survive and succeed in the Arctic (Driver 2013, 420–435). When they arrived at McComick Bay, one of the first things on their itinerary was to "search Herbert and Northumberland Islands for an Eskimo

Settlement, and if possible to induce a family to move over and settle down near Redcliffe House" (Diebitsch-Peary 1893, 39).

Ikwa and Mané, a married couple, and their two children, Anadore and Nowyahrtlik, agreed to come and support the expedition (Diebitsch-Peary 1893). Mané and Ikwa introduced the Americans to other families living in the area, and mediated some of the initial conversations between Diebitsch-Peary and those that visited her settlement. I purposefully write 'visited her' settlement: a central aspect of Diebitsch-Peary's manifest domesticity was to Americanise their winter quarters in McComick Bay and establish her dominion over their home, called Redcliffe House. McComick Bay was both the site for Diebitsch-Peary's ethnographic observations and for her identity making during her first Arctic venture. Redcliffe House, the American settlement, was the constant against which she studied and evaluated the Arctic.

As Ikwa and Mané were the first Inuit to move to McComick Bay, they were also the first Inuit Diebitsch-Peary described in *My Arctic Journal.* Diebitsch-Peary made it clear from the beginning that Ikwa and Mané were different to her and Robert; she explicitly linked Inuit to animals and nature. She wrote that "these Eskimos were the queerest, dirtiest-looking individuals I had ever seen. Clad entirely in furs, they reminded me more of monkeys than of human beings" (Diebitsch-Peary 1893, 41). In late May, several local families visited Redcliffe, and Diebitsch-Peary wrote that "the simple folk have come as heralds of the approaching spring" (1893, 151). Inuit, Diebitsch-Peary noted, were guided by the changes in weather and ice. This interconnectedness—between humans and nature, and ice in particular—is key to understanding the euro-American portrayal of the Arctic in the period of Diebitsch-Peary, and beyond (Hansson and Ryall 2018). As historian Mary Louise Pratt notably argued, there is a "dehumanizing western habit of representing other parts of the world as having no history" in the practice of exploration (Pratt 1992, 219). In Diebitsch-Peary's accounts, Inuit were written not simply as uncivilised, but outside of civilisation, culture and history, deterministically defined by the icy landscape.

Diebitsch-Peary's treatment of holidays, including birthdays and religious festivals, reveals how her ethnographic and exploration practices merged into a manifest domesticity in two main ways. Firstly, the events established unity among the American crew, and the pomp and circumstance surrounding the celebrations emphasised the Americanness of their settlement. Secondly, the holidays became a basis for comparison between

American and Inuit culture. On Thanksgiving Day, Diebitsch-Peary wrote that "all the boys wore American clothing, and the room was draped with the Stars and Stripes" (Diebitsch-Peary 1893, 83). At Christmas, she staged two dinners: one for herself and the American crew, and one where she invited Inuit. Her descriptions of their Christmas dinner are particularly revealing of Peary's ethnographic practice. She wrote that she had decided to "invite all our faithful natives to a dinner cooked by us and served at our table with our dishes. I thought it would be as much fun for us to see them eat with knife, fork, and spoon as it would be for them to do it" (Diebitsch-Peary 1893, 94–95).

The Christmas dinner company consisted of M'gipsu and Inaloo, who worked for the Americans as seamstresses, their husbands Annowkah and Ahngodegipsah and two visitors, Kudlah and Myah. Diebitsch-Peary described how the Americans gave nicknames to their guests; many were meanspirited, such as "tiresome", "misfortune" and "the villain" (Diebitsch-Peary 1893, 97). The guests were asked to 'act out' how the Americans did things. Ahngodegipsah was told to "serve the company just as he had seen Mr. Peary serve us", and M'gipsu was "told to pour the tea" for everyone (Diebitsch-Peary 1893, 97). This was clearly done for the amusement of the Americans. The portrayed reaction of Inuit to the American table setting, food selection and Christmas traditions suggested that Inuit in the area had not had any sustained contact with Europeans, and that Diebitsch-Peary's observations of their traditions were therefore representative of a more authentic indigeneity than Inuit who lived further south. This further emphasised the importance of Diebitsch-Peary's ethnographic observations and established Diebitsch-Peary identity as a white civilised woman in contrast with other women in the Arctic. Her settlement was both a place of ethnographic observation and a place where Diebitsch-Peary reinforced her manifest domesticity through an almost ritualistic performance of everyday actions (such as setting the table) and elevating holidays to anchors of civilisation.

Colonial Constructions of Ice

Holidays, in both their practice and portrayal, functioned to create a dichotomy between the known and unknown, between the portrayed civilised and the uncivilised. Ice and snow were, for Diebitsch-Peary, the key defining features of life in the Arctic. Everything else—from musical traditions and art to history and religion—was secondary and shaped by the environment. By continuously comparing Inuit to nature and by emphasising their difference to her portrayed self, Diebitsch-Peary wrote Inuit into a nature she described as fundamentally hostile. There is a tension in Diebitsch-Peary's descriptions of ice: Ice is represented as an untouched culture-free space that was available for exploration, while also being an active agent that threatened to kill her and her husband while he travelled. This is revealing of how the intersection of race, gender and colonial expansionism functioned as part of euro-American Arctic expedition.

Diebitsch-Peary was in their campsite while Robert was surveying with a small inland ice-party; he had been away for some time at this point. She was not alone, but her narrative expresses an intense sense of solitude and fear. She wrote: "The inland ice seems to wear a continual smile, so bright does its surface appear. Does it wish to assure me that all is well with the ones who are traveling on its bosom, or is it only mocking me? I will try to think the former" (1893, 171). Diebitsch-Peary expressed fears for their safety, particularly as she wrote: "Our Eskimo friends cheerfully assure us that they will never return" (1893, 158). In Diebitsch-Peary's retelling of this conversation, the local community told her that Robert and his small party were unlikely to survive their journey and that they would disappear in the ice. It is difficult to tell if the described 'cheerfulness' was a misunderstanding, or if Diebitsch-Peary was playing this up for textual effect.

Diebitsch-Peary's narrative retells her growing doubt that Robert would return, though she initially was confident in his success. Diebitsch-Peary and Robert's experience were portrayed as parallel journeys, one actual and one symbolic, which both functioned to assert their difference to the local community. While Robert overcame the danger of the ice, Diebitsch-Peary's overcame the danger of giving up hope—and of leaving behind her American identity. This perceived threat to her identity (a white American woman) and her faith that Robert would return from his expedition was emphasised in the performative descriptions of how she kept house, her body and her choice of dress. Diebitsch-Peary went through the rituals of celebrating holidays without Robert, and her room was given a "thorough cleaning" to make "things as bright and clean as possible" in the hope that it would "look somewhat homelike to Mr. Peary when he comes back" (1893, 168). Keeping one's house and body clean was, in Diebitsch-Peary's accounts, completely foreign to Inuit, and was a defining difference between herself and the other women she lived with while in the Arctic.

When Diebitsch-Peary gave birth to her daughter Marie, she further emphasised the contrast between civilised womanhood and savage maternity. Marie's nickname, the Snow Baby, was given to her because the visitors "wanted to touch her to see if she was warm and not made of snow, she was so white" (Diebitsch-Peary 1901, 17). *The Snow Baby* was also the name of Diebitsch-Peary's first children's book, with 'snow' referring both to Marie's place of birth and to the colour of her skin. Whiteness was a defining characteristic in Diebitsch-Peary's descriptions of herself and Marie. Whiteness was also associated with cleanliness and, by extension, American civilisation. Marie was bathed daily, almost ritualistically. By contrast, Diebitsch-Peary noted that Inuit children were rarely allowed in her home, "as they were not very clean" (1901, 20). When Diebitsch-Peary and Marie returned to the US, they brought a young Inuk girl E-Klay-I-Shoo (known to Euro-Americans as Bill, Billy or Billy-Bah) to stay with them in Washington for a year. "She had never had a bath", Diebitsch-Peary wrote, until she "gave her one on board the ship, and she could not understand why she must wash herself and brush her hair every morning" (1901, 39). In *Children of the Arctic*, a now older, four-year-old Marie was in the Redcliffe settlement when a few Inuit women with young children came by. Marie wanted to play with the children, and asked E-Klay-I-Shoo "to wash them tomorrow so I can play with them" (1903, 82).

Diebitsch-Peary association of American bodies with cleanliness and Inuit bodies with dirtiness demonstrates how she co-constructed herself as a unique, but authoritative, ethnographic observer while denigrating Inuit—especially Inuit women. These were racialised contrasts that were bound in an environmental determinism. Through Marie's voice, Diebitsch-Peary lamented that E-Klay-I-Shoo had not followed the rules for bathing and clothing that Diebitsch-Peary had instilled in her while she was in the US, but still "kept herself just as dirty as her companions" (1901, 57). The point here being, that Diebitsch-Peary's portrayed civilising influence had not permanently changed E-Klay-I-Shoo's behaviour, which was, fundamentally, determined by the Arctic. In referring to adult Inuit as a "child of nature" (1893, 60) and repeatedly listing how Inuit children were different to Marie, Diebitsch-Peary portrayed the icy Arctic environment as creating ways of life that were fundamentally foreign to the American ideal—an ideal Diebitsch-Peary portrayed herself as upholding in both the US and the Arctic.

Conclusion

While in the Arctic, Diebitsch-Peary photographed glaciers, hunted, cooked and made observations together with the local peoples and other foreign explorers. She collected natural history specimens and made ethnographic observations. All of these activities were included in her narrative and children's books in an informal language of female domesticity. As a lecturer and author, she took a leading role in raising money for her and Robert's Arctic explorations. Robert and his co-travellers were not able to gain the same financial and practical support from the government or scientific societies as past Arctic explorers. Thus, they had to find the funds in other ways. Lecturing and publishing were central in this fundraising effort. This is an important point, because publications like Diebitsch-Peary's narrative were also mission statements. Though they were framed as voyager accounts, travel narratives presented the explorers and their projects favourably. They could be used to showcase that these projects were worthwhile to potential supporters and investors, and that the author was a great candidate for patronage. To portray herself as someone who belonged in ventures traditionally reserved for Euro-American men, Diebitsch-Peary drew on a rhetoric of manifest domesticity. This, in turn, shaped her ethnographic practices.

The scientific contribution of Diebitsch-Peary's work drew on one of her strengths as a woman traveller. Because of her gender, she could portray herself as having privileged access to intimate aspects of Inuit culture. That is, she could undertake ethnographic observations not previously available to male ethnographers. Reviews of her book show that contemporary readers also framed it as an ethnographic work. Throughout *My Arctic Journal*—and in the advertisements for the book and her public talks—the emphasis was squarely on the supposed uniqueness of Diebitsch-Peary's experiences. This was purposefully created by Diebitsch-Peary and Robert. The construction of Diebitsch-Peary as a unique observer of Arctic culture, by virtue of her gender, took place through a racialised and imperialistic language of contrasts. These contrasts were based on Diebitsch-Peary's gendered performances of an ideal American civilisation, and encounters with all Arctic life as the other. In Diebitsch-Peary's accounts, Arctic life was fundamentally shaped by nature, and her labour upheld American domesticity in the face of a hostile icy landscape.

In negotiating the traditionally masculine tropes of Euro-American Arctic exploration, Diebitsch-Peary created a narrative voice that drew on the literary techniques and genres used by many women science writers in the nineteenth century. Diebitsch-Peary did not elaborate on ethnographic theory in *My Arctic Journal*, and she moved fully into a maternal narrative voice with her two children's books. Yet, her work should be seen as an early and significant example of women's fieldwork-based ethnographic research. Diebitsch-Peary's books were widely read, and her lectures were very popular. The way she portrayed Inuit should therefore be taken seriously when considering how imperialistic ideas about Arctic Indigenous peoples were created and upheld in the US. For Diebitsch-Peary, life in the Arctic was summarised by the sub-title of her narrative: *A Year Among Ice-Fields and Eskimos*. Diebitsch-Peary linked Inuit to ice in a form of colonial violence coated in the language of polite American society and science.

Bibliography

Aarekol, Lena. 2020. The Snow Baby Books Mediating Arctic Experiences to Children. In *The Arctic in Literature for Children and Young Adults*, ed. Heidi Hansson, Maria Leavenworth, and Anka Ryall, 57–70. Abingdon-on-Thames: Routledge.

Alacovska, Ana. 2015. Genre Anxiety: Women Travel Writers' Experience of Work. *The Sociological Review* 63: 128–143. https://doi.org/10.1111/1467-954X.12246.

Anon. 1884. Mrs. Peary at the Pole. *Freeland Tribune 3*, 22 January.

———. 1891. A Woman as an Explorer. *The Morning Call 7*, 21 June.

———. 1892. Keeping House in the Arctics. *Pittsburgh Dispatch 9*, 13 November.

———. 1892. Mrs. Peary. *The Iola Register*, December 2.

———. 1893. The Gossip of Gotham. *The Indianapolis Journal 4*, 12 October.

———. 1893. The Gossip of Gotham. *The Morning News 12*, 8 October.

———. 1893. The Gossip of Gotham. *Wheeling Register 8(9)*, 8 October.

———. 1893. *The New-York Tribune 5*, 29 January.

———. 1893. *The Times 3*, 24 November.

———. 1894. *The Salt Lake Herald 3*, 21 January.

———. 1894. *The Scottish Geographical Magazine 152*, 1 March.

———. 1902. Peary's Expeditions. *New-York Daily Tribune 2*, 19 September.

———. 1909. Mrs. Peary Expected It. *New-York Daily Tribune 2*, 7 September.

———. 1909. Mrs. Robert Edwin Peary, the Wife of the Polar Explorer. *The Sunday Star 3*, 21 November.

Barth, Fredrik, Andre Gingrich, Robert Parkin, and Sydel Silverman. 2005. *One Discipline, Four Ways: British, German, French, and American Anthropology*. Chicago/London: University of Chicago Press.

Bergmann, Linda. 1993. Women Against a Background of White: The Representation of Self and Nature in Women's Arctic Narratives. *American Studies* 34: 53–68.
Bloom, Lisa. 1993. *Gender on Ice: American Ideologies of Polar Expeditions.* Minnesota: University of Minnesota Press.
Browne, Janet. 1994. A Glimpse of Petticoats: Women in the Early Years of the British Association for the Advancement of Science. In *Science and British Culture in the 1830s*. Cambridge: Conference Proceedings, Trinity College.
David, Robert G. 2000. *The Arctic in the British Imagination 1818–1914.* Manchester: Manchester University Press.
Diebitsch-Peary, Josephine. 1893. *My Arctic Journal: A Year Among Ice-Fields and Eskimos.* New York/Philadelphia: Contemporary Publishing Co.
———. 1901. *The Snow Baby: A True Story with True Pictures.* New York: Frederick A. Stokes Company.
Diebitsch-Peary, Josephine, and Marie Ahnighito Peary. 1903. *Children of the Arctic.* New York: Frederick A. Stokes Company.
Driver, Felix. 2013. Hidden Histories Made Visible? *Transactions of the Institute of British Geographers* 38 (3): 420–435.
Erikson, Patricia. 2013. Homemaking, Snowbabies, and the Search for the North Pole: Josephine Diebitsch-Peary and the Making of National History. In *North by Degree: New Perspectives on Arctic Exploration*, ed. Susan A. Kaplan and Robert McCracken Peck, 254–285. Philadelphia: American Philosophical Society.
Grace, Sherrill E. 1997. Gendering Northern Narrative. In *Echoing Silence: Essays on Arctic Narrative*, ed. John George Moss, 163–181. Ottawa: University of Ottawa Press.
Hanrahan, Maura. 2017. Enduring Polar Explorers' Arctic Imaginaries and the Promotion of Neoliberalism and Colonialism in Modern Greenland. *Polar Geography* 40: 102–120. https://doi.org/10.1080/1088937X.2017.1303754.
Hansson, Heidi. 2009. Feminine Poles: Josephine Diebitsch-Peary's and Jennie Darlington's Polar Narratives. In *Cold Matters: Cultural Perceptions of Snow, Ice and Cold*, ed. Heidi Hansson and Cathrine Norberg, 105–123. Umeå: Umeå University and the Royal Skyttean Society.
Hansson, Heidi, and Cathrine Norberg, eds. 2009. *Cold Matters: Cultural Perceptions of Snow, Ice and Cold.* Umeå: Umeå University and the Royal Skyttean Society.
Hansson, Heidi, and Anka Ryall, eds. 2018. *Arctic Modernities: The Environmental, the Exotic and the Everyday.* Cambridge: Cambridge Scholars Publishing.
Hansson, Heidi, Maria Leavenworth, and Anka Ryall, eds. 2020. *The Arctic in Literature for Children and Young Adults.* Abingdon-on-Thames: Routledge.
Harper, Kenn. 1986. *Give Me My Father's Body: The Life of Minik, the New York Eskimo.* Frobisher Bay: Blacklead Books.
———. 2002. The Minik Affair: The Role of the American Museum of Natural History. *Polar Geography* 26: 39–52. https://doi.org/10.1080/789609352.

Heggie, Vanessa. 2019. *Higher and Colder: A History of Extreme Physiology and Exploration*. Chicago: University of Chicago Press.
Hill, Jen. 2009. *White Horizon: The Arctic in the Nineteenth-Century British Imagination*. Albany: State University of New York Press.
Kaplan, Amy. 1998. Manifest Domesticity. *American Literature* 70: 581–606.
Larsen, Kristine. 2017. *The Women Who Popularized Geology in the 19th Century*. Cham: Springer.
Larson, Frances. 2011. 'Did He Ever Darn His Stockings?' Beatrice Blackwood and the Ethnographic Authority of Bronislaw Malinowski. *History and Anthropology* 22: 75–92. https://doi.org/10.1080/02757206.2010.487869.
Lightman, Bernard. 2009. *Victorian Popularizers of Science: Designing Nature for New Audiences*. Chicago/London: Chicago University Press.
Lind, Mathilde Frances. 2020. Her Unwritten Legacy Women's Projects and the History of Folklore Studies. *Western Folklore* 79: 401–426.
Menand, Louis. 2001–2002. Morton, Agassiz, and the Origins of Scientific Racism in the United States. *The Journal of Blacks in Higher Education* 34: 110–113.
Pratt, Mary Louise. 1992. *Imperial Eyes: Travel Writing and Transculturation*. London/New York: Routledge.
Reeploeg, Silke. 2019. Women in the Arctic: Gendering Coloniality in travel Narratives from the Far North, 1907–1930. *Scandinavian Studies* 91: 182–204. https://doi.org/10.5406/scanstud.91.1-2.0182.
———. 2021. Gendering Arctic Memory: Understanding the Legacy of Josephine Diebitsch-Peary. *Memory Studies* 14: 1061–1080. https://doi.org/10.1177/17506980211024327.
Robinson, Michael. 2015. Manliness and Exploration: The Discovery of the North Pole. *Osiris* 30: 89–109. https://doi.org/10.1086/682968.
Rud, Søren. 2013. Erobringen af Grønland: Opdagelsesrejser, etnologi og forstanderskab i attenhundredetallet. *Historisk Tidsskrift* 106: 488–520.
———. 2017. Ethnography, Time, and the Idealization of Tradition. In *Colonialism in Greenland: Tradition, Governance and Legacy*, ed. Søren Rud, 9–31. Cham: Springer International Publishing.
Sera-Shriar, Efram. 2013. *The Making of British Anthropology 1813–1871*. London/Pittsburgh: Pickering and Chatto/University of Pittsburgh Press.
Shteir, Ann. 1997. Elegant Recreations? Configuring Science Writing for Women. In *Victorian Science in Context*, ed. Bernard Lightman, 236–255. Chicago/London: University of Chicago Press.
Vider, Janikaa. 2021. Lady Scientist, Technological Treasures and the Making of a Discipline Re-assembling the Histories of the Czaplicka Collection at the Pitt Rivers Museum. *Journal of the History of Collections* 33: 111–127.
Visweswaran, Kamala. 1997. Histories of Feminist Ethnography. *Annual Review of Anthropology* 26: 591–621. https://doi.org/10.1146/annurev.anthro.26.1.591.

Materiality of Time: Polar Ice as a Medium for Ecological Art for the Tempered Zones

Birgit Schneider

My Polar imaginary was recently inspired by an inconspicuous, modest object I discovered in an exhibition drawer in the Polar Museum in Cambridge, UK. This museum was founded in 1920 as a national memorial to Robert Falcon Scott and his companions, who died during their journey to the South Pole in 1912. The drawer contained a flattened, light brown paper cookie wrapper left over from the fatal expedition. The curators of the exhibitions found this object—a piece of leftover garbage from an empty cookie pack—worthy of being displayed, without further comment. To me, it became a minimalistic object that highlighted death and failure as the flip side to heroic expeditions, much like global warming is the flip side of the great successes of an energy-intensive lifestyle.

The regions north of the Arctic Circle had been explored for centuries, especially during times of European expansionism since the Renaissance. Since the end of the nineteenth century—when James Cook and his crew

B. Schneider (✉)
Potsdam University Institute for Arts and Media, Potsdam, Germany
e-mail: birgit.schneider@uni-potsdam.de

A. Hemkendreis, A.-S. Jürgens (eds.), *Communicating Ice through Popular Art and Aesthetics*, Palgrave Studies in Media and Environmental Communication,
https://doi.org/10.1007/978-3-031-39787-5_6

crossed the Southern Polar Circle for the first time in recorded history—heroic expeditions to Antarctica were in the name of science. With the rising notion of global warming, the poles again became a spot for scientific exploration, and, simultaneously, symbolic regions for communicating climate change. Since the turn of the millennium, more and more artists have gone to the poles following the idea that in times of global warming you have to go to the distant (Ant)arctic regions to be able to share the experience in temperate latitudes, because artists have a special ability to make distant and abstract themes aesthetically comprehensible and tangible. Thus, ice not only functions as a trope or metaphor, but is increasingly an active agent of scientific and artistic communication about the melting poles.

The Arctic sublime emerged as a reaction to nineteenth-century expeditions, as an aesthetic way to imagine and to communicate the Polar regions. Chauncey C. Loomis described how the landscapes "haunted the Victorian imagination" of the "armchair travellers" (1977, 96). Within "the cold vastness and indifferent powers of the inorganic cosmos", the "Natural Sublime" evoked "a mixed sense of triumph and defeat", of simultaneous "mystery" and "terror" (Loomis 1977, 96, 98 and 104). In the nineteenth century, the Polar regions were zones of the uncanny and even terrifying, because they caused the death of early expeditioners; today these zones again evoke feelings of the uncanny and the catastrophic, because they produce vast and immeasurable phenomena—such as decreasing stability of ice sheets and high-speed melting events caused by climate change. In times of declining ice—due to rising temperatures and changing climates—the old tropes of the Polar regions and the "arctic sublime" (Loomis 1977) still feed the imagination, but the sublime gains a new layer of meaning, because it is neither eternal nor perceived from a secure position. The idea of 'eternal ice' must give way to the idea of transience and radical change. The aesthetic perception of the sublime today is inextricably linked to the vastness of the climate crisis. Building on this mythical and epical imagination—still shaping today's imaginations—the editors of this volume at hand update the meaning of Polar ice as a "figure of existential crisis" and "a sign of irretrievable loss", because it signals the destruction of entire ecosystems and even mankind itself (quoted from the abstract to the conference "Ice(St)Ages" 2021).

Melting ice is the most evident material for making global warming tangible. It symbolises, advocates and proxies global warming. Calving glaciers, breaking ice floes, floating icebergs and rising meltwater are visible traces and signs of rising temperatures. Climate-change-induced temperature increases are not distributed evenly over the globe; temperatures are, and have been, rising most dramatically above the poles, making them 'hot spots' and early harbingers of global warming. Due to the Polar regions' vulnerability, and their distance to most people's experience, the poles visualise climate change consciousness for those in other climate zones; they became areas where climate change is witness-able. Scientists bring home observational data, abstract measurements or many kilometres of drilled ice cores—which are tough to communicate to a 'general public'. In contrast, artists take the role of outreach staff for what is going on at the poles. They return with different forms of testimony: aesthetic perceptions, new imaginaries and stories of changing landscapes.

In the following, I reflect on ice as a medium for art and imagination. I emphasise and elaborate on the ambivalence of artist expeditions, which attempt to raise global awareness about the climate crisis in industrialised and tempered zones, but are in danger of slipping into old patterns of imperial voyages of discovery. Only, this time, it is not the 'untouched' worlds that are marvelled at, but the remote impacts of industrial society's way of life. It is debatable to what extent the (well-intentioned) artists are repeating an imperial gesture, even though it is framed as a discovery and testimony of a world declining. This is all the more poignant because the effects they observe at the poles are the consequences of their own lifestyle and home country's socio-political history.

Starting from an elementary perspective on the medium of ice, I focus on works of art from 2001 to 2017—shortly after the success of the fifteenth Conference of the Parties (COP15) and Paris Agreement—that seek to evoke icy imaginaries in temperate-zone museums. In terms of an "ecologically motivated aesthetic" (Böhme 1995, 8), art allows us to delve into exactly what ice allows us to sensualise and imagine, but also problematises the artistic mode of sublimity in times of global warming, and art's connection to Polar expeditions. Examples include artworks by Hans Haacke, David Buckland, Katie Paterson, Olafur Eliason, Erika Blumenfeld, Julian Charrière and Mathias Kessler (among others), but also the Antarctic Biennale in 2017. The chosen artworks exemplify general perspectives on ice; since the millennial turn there is an ever-growing group of works which might be termed "Freezing Art" (Wichmann 2021). I take a Western perspective, which is rooted in the 'tempered zones' and is very different to the one of the people of the Polar Regions—who have had to fight for "the right to be cold", to protect their cultures, since the 1980s (Watt-Cloutier 2015).

Ice as a Medium in Climate Aesthetics and Icy Imaginaries

The editors of this volume phenomenologically and culturally identify ice as a medium which allows experiencing loss but also the mythical dimensions of eternal ice (quoted from the abstract to the conference "Ice(St) Ages" 2021). If we consider (Polar) ice to be a natural or elemental medium, what would this phenomenon signify? What meaning does it contain? What is the index of (Polar) ice and icy landscapes elsewhere?

Much has been written in media theory about natural mediums—such as air, water or clouds—although ice has not yet had much attention as a medium. Clouds, for example, are especially interesting to media theory, because of their ability to appear and disappear, and their bizarre and manifold shapes, which are simultaneously formed and formless; this is why clouds offer a perfect starting point to reflect about ephemeral media (as Aristotle did). While clouds have been differentiated by a taxonomy (e.g. cumulus, cirrostratus), such names for the different shapes of the cryosphere do not exist; if ice is differentiated it is into objects such as ice shelves, ice sheets, shelf ice or glaciers. Yet, many of the form(less) aspects of clouds also apply to ice. Ice is the solid state of water; although it is often unperceivable to the human eye, it does move and change (slowly). On a connotational layer ice mediates permanent change and passing time on different scales: moments, hours, days, months, years, decades or even geological time scales (e.g. hundreds of thousands to millions of years). Despite being an agile and ephemeral material, ice is normally perceived as inorganic; it seems neutral to the processes that change it so dramatically. Even though it is an actor and a stage for other actors—like penguins in the Antarctic and cultures in the Arctic region—undergoing radical transformation, it does not care about its state of aggregation.

If ice is a medium and media are environments, as Marshall McLuhan wrote, this idea is productive for exploring icy landscapes (McLuhan 1969, 31). As material, ice archives spatiotemporal conditions and, consequently, information. It is a geological material and formation—a medium which carries specific semantics and shows "nature herself an artist, the force behind all forms" (Böhme 2016, 37).

Ice as a Medium in Art and as *Real-Time System*

Independent of the Arctic Sublime, ice has its own significance as a material and natural medium in Western art. Ice is put to artistic use and explored in artistic ways because melting ice is a well-known experience for all people in the temperate zones, and it is the direct reflection of rising temperatures. Therefore, "existential experiences of estrangement and loss of words in the face of looming disaster" (quoted from the abstract to the conference "Ice(St)Ages") can be related to ice experiences, regardless of whether these experiences are mediated or made directly on the material.

On a phenomenological layer, ice is a physical experience of the planetary element of water. Water is a material which impressively changes state in relation to temperature and pressure: from liquid to solid and solid to liquid; from rain and aerosols (i.e. fog, steam, water vapour, clouds) to ice or crystalline snow. Because ice is a "Realzeitsystem"—a *real-time system*, as German artist Hans Haacke termed it in the late 1960s—it offers a specific materiality of time (Fry 1972, 10).

The medium of fluid water has art historically been framed in fountains and gardens. This is where water got highlighted as an eternal element of life: permanent flow and nourishment. In contrast to this water in flux, ice is a standstill and was not used in art as a material until the modern art era of the 1960s. Before this, ice was fascinating as a creation of nature (*natura naturans*)—a natural power which stands behind all forms—but if it became an object of art, it was treated as a representation only. Most known are the many Flemish examples showing icy landscapes and people interacting with these landscapes during the Little Ice Age in the sixteenth and seventeenth centuries. The most famous is probably the *Winter Landscape with Ice Skaters and Bird Trap* by Pieter Bruegel the Elder from 1565 (Royal Museums of Fine Arts of Belgium), in which ice is a passive stage for human activity. While the mentioned examples are an account of icy landscapes in Europe, the fix point of the imagination of an Arctic Sublime is Caspar David Friedrich's *The Sea of Ice* from 1823 to 1824 (Hamburger Kunsthalle), which Friedrich painted after reading accounts of expeditions to the North Pole. Although there are many other images depicting ships wrecked in ice from the nineteenth century, this iconic picture emphasised the sublime idea of natural forces being superior to humans (Potter 2007) by showing the shipwreck crushed only at second glance in the background of massive ice floes. We will see how, even still,

Friedrich's image layout is a fix point for artists dealing with climate change and icy imaginaries.

To find works of art that not only depict ice but also integrate it as a material and natural phenomenon, we have to make a big leap in time to the modern era. German artist Hans Haacke (*1936) was one of the early artists who used ice as a medium for many of his early sculptures. He is of particular interest for this chapter because he coined the idea of art as a "real-time system", which can be observed in the process (Fry 1972, 10). For a 1971 solo exhibition planned for the Guggenheim Museum (New York)—but which was cancelled because he refused to withdraw two political works—Haacke had prepared several works around the hydrological cycle and the changing states of H_2O. Most famous is Haacke's *Kondensationswürfel* (Condensation Cube), which he built for the first time in 1963 and which changes in appearance depending on the climatic conditions of the museum space. The cube is made of "acrylic glass, water, light, air streams, environmental temperature", as Haacke subtitles the work (DuMont 1972). Curator Edward Fry related Haacke's weather works to the idea of the ready-made in the 1972 German catalogue:

> By creating the conditions under which water can evaporate and condense in response to light changes, temperature changes, and air currents in the environment, [Haacke] ultimately put his name to a universal natural phenomenon, singling it out but still allowing it to function normally – independent of further human intervention and by virtue of its own existence. The weather boxes, as Haacke aptly called them, thus extended Duchamp's principle of the ready-made to the at least potential inclusion of all real world phenomena: all its result of what the artist may select to 'articulate something natural'. (Fry 1972, 11, translation B.S.)

The works included in the 1972 monograph demonstrate Haacke's intense preoccupation and fascination with the natural process of freezing and melting. There are eight works which included ice as process and material: *Floating Ice Ring* with a cooling unit on acrylic (1970); an *Ice Bar*, mounted on a pedestal and consisting of a cooling unit with an electric regulator (1966); a square *Ice Table* consisting of a cooling unit (1967). The concept of ice or snow as real-time system becomes especially vivid in five photographically recorded works: Schneehaufen schmelzend und verdunstend, 10, 11, 12 February 1969, Dach von 95 East Houston Street, New York (*Snow Piles Melting and Evaporating, Feb. 10, 11, 12, 1969, on the roof of 95 East Houston Street, New York*); Gegossenes Eis gefrierend und schmelzend, 3–5 January 1969, Dach von 95 East Houston Street, New York (*Poured ice freezing and melting, Jan. 3–5, 1969, on the roof of 95 East Houston Street, New York*); Graben im Schnee, parallel zur von den Gezeiten bestimmten Schneegrenze, 1969, Coney Island (*A Trench in the snow, parallel to the tide-determined snow line, 1969, Coney Island*) (Fig. 1); Eisscheibe in gefrorener Umgebung, 16. Dezember 1968, Dach von 95 East Houston Street, New York (*Ice disk in frozen environment, 16 December 1968, on the roof of 95 East Houston Street, New York*); Sprühregen der Ithacafälle auf einem Seil gefrierend und schmelzend, 8. February 1969 (*Spraying rain from Ithaca Falls freezing and melting on a rope, February 8, 1969*).

Haacke's works from the late 1960s did not relate to climate change, because back then ecological discourses were led by other themes, such as

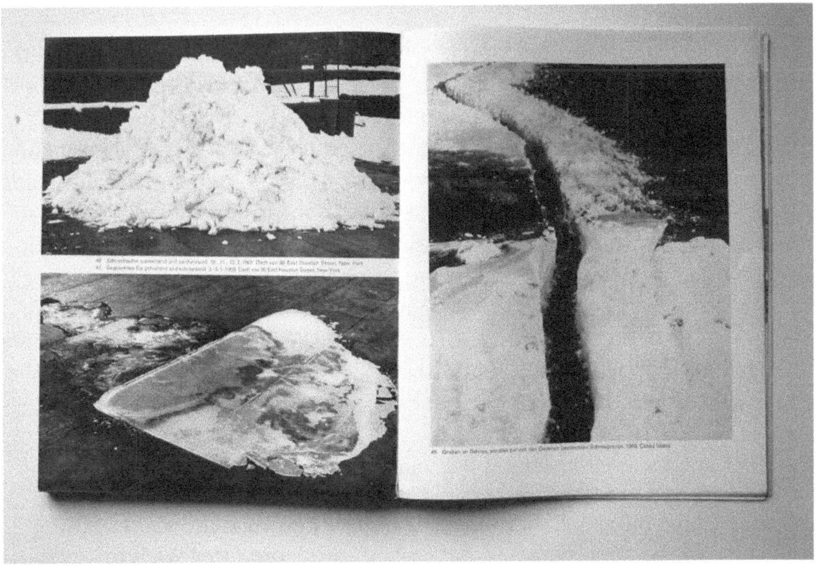

Fig. 1 Hans Haacke, Werkmonographie, 1972. (From left to right: snow piles melting and evaporating on the roof of 95 East Houston Street, New York in February 1969, poured ice freezing and melting on the same roof in Jan. 1969, trench in the snow, parallel to the tide-determined snow line, on Coney Island in 1969)

pollution and exploitation. Instead, Haacke took ideas from system theory and cybernetics, which were shaping ecological thinking and the emerging notion of the biosphere at the time, and processed them further within his art. In the context of anthropogenic climate change, water and ice as elemental media for art take on a different connotation, but still rely on the basic metamorphoses and process thinking which Haacke made tangible in his early ice works.

Artistic Travellers Bringing Icy Experiences to the Temperate Zones

In the following I analyse different contemporary art works which deal with ice from the poles and contrast human time against geological time. Most also consist of representations and mediations that carry icy imaginaries to the temperate latitudes, but they were created through direct contact with the ice at the poles, thus claiming authority based on experience.

British artist David Buckland (*1949) may have been the first to call for artists to embark into the Arctic Sea on research vessels in reaction to the climate crisis. He started the international *Cape Farewell* charity project in 2001, which brought together scientists, educators, artists, explorers and creatives since then to explore new lines of research and work alongside one another to create interdisciplinary work informed by 'first-hand' scientific research. In doing so, *Cape Farewell* wanted to shift the perception of climate change and get the attention of a larger audience, because 'climate is culture', as the guiding slogan of *Cape Farewell* tells us. Buckland states that the Arctic is a place to witness the losses caused by climate change and that artists can be translators for climate science to the public, because they can make the abstract concrete on a human level.

> The Cape Farewell invitation was open: come to the Arctic, engage with scientists and, we hope, be inspired to make art. Art is never a guarantee; many a time I have followed a path of enquiry only to abandon it in frustration. What is extraordinary is that each artist has produced work that celebrates authorship. When unpacked, each artist has in some form responded to this cold, Arctic place and the way it is changing as the Earth warms. They have all told the story on a human, rather a planetary scale. (Buckland 2006, 6)

For his works in the early 2000s, Buckland took photographs of blue icy landscapes to increase awareness of global warming; short statements like "BURNING ICE", "ANOTHER WORLD IS POSSIBLE" or "DISCOUNTING THE FUTURE" were projected in white capital letters from the expedition ship onto the bluish bases of crumbling icebergs. The texts induce emotional responses due to the haunting lowlight and sombre tone, but they can also be understood as a protest statement. Buckland commented on his understanding of ice as a semantic medium by saying "Ice is alive. […] It has a language that is as clear as words, and understanding it is like dealing with poetry and raw emotion. […] Filming the demise of an iceberg is both exhilarating and sad. As we watch it sink lower and lower, and finally collapse, our reflection turns towards the implication of the loss of ice" (Buckland quoted from Brown 2014, 77). For Buckland, the Arctic icebergs are like patients in a palliative care unit.

Scottish artist Katie Paterson (*1981) also went where there is perpetual ice—the glaciers of Iceland. In 2007 she recorded sounds of three melting glaciers: Langjökull, Snæfellsjökull and Solheimajökull. The recordings were pressed into three LPs with sound grooves, then cast and frozen using the meltwater from each corresponding glacier. For Paterson's artwork the frozen disks are important as a real-time system in the sense of Haacke, too: The discs of ice were then played simultaneously on three turntables until they melted completely. Paterson filmed this process and exhibited the three videos in an installation. Visitors thus witness the records melting before their very eyes, including how this slowly and audibly destroys the condition of the sound. The idea of letting people in temperate latitudes explicitly experience the melting of Icelandic glaciers was even more directly realised in another work in which the artist made "those temporal and spatial dimensions that eschew human comprehension both tangible and visible" (Holzhey 2016, 124). A live phone line was created to an Icelandic glacier, via an underwater microphone submerged in Jökulsárlón lagoon, an outlet of Vatnajökull. The number 07757001122 could be called from any telephone in the world, and the listener would hear the sound of the glacier melting—like pressing a stethoscope onto the chest of a patient who is slowly passing away.

Danish and Icelandic artist Olafur Eliasson (*1967) travelled to icy regions to bring back many tons of authentic ice from the Greenlandic ice sheet for his 2014 work *Ice Watch*. Haacke's approach to the real-time system is at the centre of this work. Twelve (in later installations twenty-four) large blocks of ice that had calved from the Greenlandic ice sheet were harvested from a fjord outside Nuuk and brought to the centre of Copenhagen. Eliasson, in cooperation with Greenlandic geologist Minik Rosing, installed *Ice Watch* at different city hall squares—like Copenhagen, London or Paris—to mark the publication of the United Nation IPCC's *Fifth Assessment Report on Climate Change*. A hot spot of climate change is brought to the European cities, so people can bear witness to the melting of the Arctic right there with their own senses. In the many photos on Eliasson's website, one can trace the intense contact people made with the strange ice blocks from the far north. Like Caravaggio's *Doubting Thomas* from 1601 to 1602—the apostle who refused to believe without direct personal experience—people can touch the evidence with their own fingers. This is the highly didactic layer of Eliasson's piece, which might be seen as an overwhelming and simple gesture of concern for people. Then again, it is interesting to see that visitors really want to feel, sense and interact with the Arctic ice; they touch, smell and taste the ice, some even crawl into the molten voids, as they want to get as close to the ice as possible. Even though the piece is ambivalent because of its spectacular impression (but also because it raises the question of appropriation of Greenlandic ice by a European artist), it successfully connects the icy imaginary of the temperate zones to personal experience in order to create attention and concern in times of global warming. "Thus, they become objects of a yearning for an untouched, constant nature, which results in a neo-romantic longing and accordingly defines mankind's relationship to nature as well as its paradigms of action" (Hosseini, 2019). The marketplaces of human politics are set against the distant poles. The iceberg is brought to parliament to speak for itself. The clock ticking down seems to say 'Make politics in the name of the melting poles!'

Swiss artist Julian Charrière's 2013 series *The Blue Fossil Entropic Stories* can be contrasted with Buckland's photographs. Charrière (*1987) went to Iceland, too, but he embarked on the Arctic Sea close to Iceland. For *The Blue Fossil Entropic Stories*, he brought home a series of photographic traces of a visit to a small iceberg floating in the sea (Fig. 2). The actions he inflicted on the iceberg follow the tactic of overidentification, exaggeration and subversive affirmation. After he was dropped off on the iceberg like a mad explorer he started melting the ice for eight hours with a gas torch. "Like an absurd, quixotic hero, Julian Charrière confronts the elements in a seemingly hopeless battle—human time against geological time. And yet, a battle of which global warming is only the starting point" (Charrière 2023). A lonely man fighting nature. Charrière, who

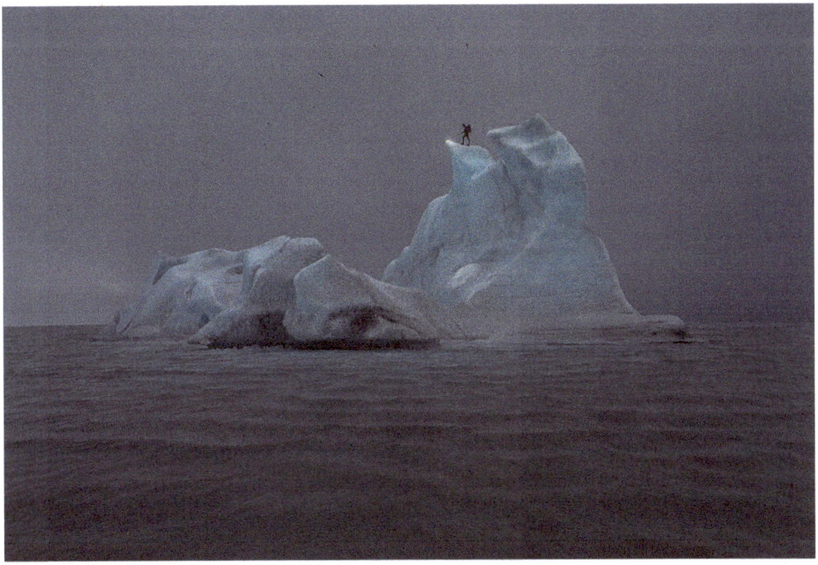

Fig. 2 Julian Charrière, *The Blue Fossil Entropic Stories* (3), 2013 © the artist; VG Bild-Kunst, Bonn (Germany)

intensively uses the magnificent beauty of images and their sublime cruelty, compared his image to Caspar David Friedrich's picture *Wanderer Overlooking the Sea Fog* (1817–1818) (Charrière 2023). Immanuel Kant defined the sublime as an experience of a fearful nature despite knowing oneself to be in a safe position. "Yet the sight of them becomes all the more attractive the more fearful it is, provided we are in a safe place" (Kant 1987, § 28). But, in this case, the lonely wanderer is not unrelated to what he looks at; he actively destroys the ground on which he stands together, along with the possibility of contemplating the sublime from a position of safety. Charrière, as an individual, symbolically embodies the destructive activity for all societies following high-carbon lifestyles.

Bringing Art to the Pole: The Antarctic Biennale

The aforementioned artists went to the Arctic to bring home artworks which made their experiences tangible. There was also an opposing movement, which brought art to the "eternal ice": *The Antarctic Pavilion* presented the concept of an *Antarctic Biennale* at the *Venice Biennale*. The *Antarctic Biennale* eventually took place in 2017, and actually happened in Antarctica.

Artist and former Russian naval officer Alexander Ponomarev and curator Nadim Samman are responsible for both undertakings. They write on their website that the *Antarctic Pavilion* "propose[s] an expanded Antarctic imaginary" (Antarctic Pavilion 2015). They also allude to the Western icy imaginary of an untouched wilderness—of "all the kilometres of ground untrodden by human foot, mountains unnamed and creatures unknown"—and claim Antarctica as a cultural space (Samman 2015; Antarctic Pavilion 2015). At the same time, the transnationality of the Pavilion "constitutes a polemical engagement with the Biennale's nationally over-determined structure: a quasi-institutional claim to represent a transnational sphere, out of line with the festival's politics of territorial representation" (Antarctic Pavilion 2015). But actually, the Antarctic Pavilion is a European starting point for a biennale to be held in Antarctica.

For their 'Antarctic Biennale expedition' in March 2017, Ponomarev and Samman invited a group of 80 people—artists, philosophers, scientists, journalists and art critics—to embark to "the only continent without a biennale" (Antarctic Pavilion 2015) on the Russian research vessel Akademik Ioffe for an 11-day voyage. Similar to *Cape Farewell*, but even more concentrated, the journey brought several performances and installations to the eternal ice. On the ship were artists such as Tomás Saraceno, Julius von Bismarck, Lara Favaretto, Julian Charrière and Andrey Kuzkin. Artist Yto Barrada laid out large, self-dyed rectangular pieces of fabric on the snow during the trip. Many artists created works critical of Western appropriation and commodification of Antarctica, and of the ambivalence of the Western perspective and myth of 'discovery'.

The images that the organisers of the Biennale use to promote their project show classic expedition views: the bow of a research vessel pushing through the sea towards an icy coast, or the artist and organiser of the biennial (Ponomarev) as he points into an endless icy landscape, bearded and dressed in expedition clothing. American art critic and philosopher Dehlia Hannah, who actually was participating in the journey, critically commented on this undertaking: "By compounding the aesthetic force of Antarctica's stark landscape, the Antarctic Biennale only served to intensify this desire for a collective experience amongst those assembled to witness this exploratory experiment" (Hannah 2017, 42). She also observed how the "participants felt the pressure to justify the apparent decadence of this 'existential cruise'" and concluded that "Antarctic culture issues from the sedimented traces left on the landscape and in the global cultural imagination by colonial explorers, whalers, scientists, tourists and the handful of artists that have previously joined research missions to the continent" (Hannah 2017, 43).

Therefore, the travelling artists embarking on the research vessel Akademik Ioffe have to ask themselves, what distinguishes them from other tourists and what they might share in their intention? Because the disappearing landscapes of ice lead to nostalgic tourism: while colonialism has since its beginning led to the demise of cultures and environments, global warming is the updated form of this fatally destructive relationship. Rich people embark on Arctic Discovery Cruises or Antarctica Luxury Cruises to witness 'the most unspoilt and remote place left on Earth'. If you can afford a ticket for the Silver Explorer, which is owned by the U.S. company Royal Caribbean Group, you are promised to "set foot on this last continent, a place untouched by humanity—and enjoy what is probably the most southernmost afternoon tea in the world" (Silversea 2022).

Taking Critical Care of Icy Imaginaries

Traveller Alexander von Humboldt—an ongoing model for ecological ways of perceiving the world—knew the value art brought to imaginaries of nature. While scientists can only bring data and measurements or classified materials, artists can take the imaginary with them. In times of climate crisis, this idea is connected to the strong belief that art can play an important role in raising awareness—in making the unthinkable imaginable. David Buckland from *Cape Farewell* cites a famous quote from McLuhan to legitimise art's role as a central transmitter of climate catastrophe and urgency to act: "I think of art, at its most significant, as a DEW line, a Distant Early Warning system that can always be relied on to tell the old culture what is beginning to happen to it" (McLuhan 1964, n. p.). This is a common assessment and expectation towards the power and potentials of art which gains an important meaning in times of global warming. It considers artists—without being able to explain this ability—to embody a prophetic access to the future similar to an early seismograph for forthcoming changes.

But what about the aesthetics of the sublime today, to what extent is it still an appropriate mode for conveying the consequences of climate change, and to what extent is it outdated? Austrian writer and poet Adalbert Stifter (1805–1868) wrote about the sight of the starlit, nocturnal sky as an experience in which nature creates spiritual affects: "And even if one cannot imagine it, there rises a beauty before us, which enraptures and makes us shudder, which delights us and destroys us. [...] There human thinking and human imagination comes to an end" (Stifter 1937, 228). Some impressions of nature cause sublime perceptions more than others. This effect seems to also be true for the Polar regions, which (again and again) are framed as boundary perceptions that transcend the imagination. In times of global warming, Arctic and Antarctic imaginaries are deeply connected to the longing of going there and having an experience that is strange, and even shocking, to people from other climate zones. Today, this impression is all the more important, because what global warming does to the poles (and to other regions) is even more unimaginable. At the same time, it is the idea of the sublime which drives this longing for visiting icy landscapes today, which no mediation seems to be able to transport. "Any Antarctic experience is a journey. You have to go there and come back" as literature scholar Elisabeth Leane—who studies the fascinating history of the Polar region and the idea that Antarctica has been represented as an "unwritable" empty ground—sums up (Leane 2012, 1). The notion that all imaginations, discussed so far, are the result of journeys to the poles goes together with the idea of this landscape as unrepresentable—or represented with an empty (white) canvas, a medium without form.

But the sublime is an ambivalent aesthetic experience in times of global warming. Austrian artist Mathias Kessler (*1968), who apparently stayed home for this artwork (although he had travelled to Greenland for his series of iceberg photographs in 2007, too), seems to allude to the ancestor of climate art in museums—Hans Haacke—when he examined and staged nature as a projection in a refrigerator labelled *The Sea of Ice—The Failed Hope* (Fig. 3). The art installation again refers to Caspar David Friedrich's famous painting *The Sea of Ice*. His installation is a real-time system, too, powered by the mundane, energy-consuming cooling device, which adds to the reasons why sublimity as an aesthetic experience is contested today. There is no safe place anymore to watch nature. The fate of the melting ice is our own fate. Kessler's increasingly icy model becomes a caricature of the longing for an untouched romantic and sublime landscape, "a sort of psychological-romantic paradox, linking the imaginary and its representation to humanity's real impact on the earth" (Vienat

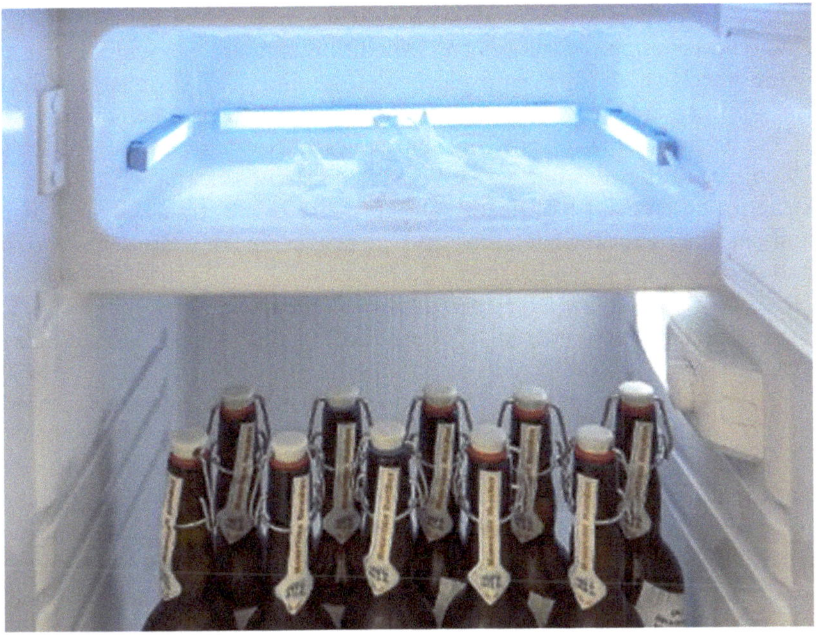

Fig. 3 Mathias Kessler, *Das Eismeer—Die gescheiterte Hoffnung* ("The Sea of Ice—Failed Hope") 2012, 3d model in mini fridge with freezer compartment © Kirchner Museum

2018). Mathias Kessler planned his installation as a "social sculpture" ("Soziale Plastik", concept by Joseph Beuys), as the beer-drinking visitors influence the icing process in the refrigerator. It is not just inhuman forces that shape the icy imaginaries, but human forces, too. The installation exposes, and deconstructs, the imaginary that remains dormant in Western museum-goers to this day.

Art historian Amanda Boetzkes recently asked, "How can a planetary sensibility be forged from a response-ability (to use Haraway's term) for the roots of climate crisis in the imperial and industrial episteme? And how might this response-ability be attuned to ongoing Indigenous political resistance, and an ethics of allyship and environmental stewardship?" (2020, 38). Therefore, I want to resume: I started this chapter with a flattened, light brown paper cookie wrapper left over from the fatal expedition of Robert Falcon Scott and his companions in 1920. This piece of rubbish, which is trivial and even ridiculous, was left behind in the Antarctica as a trace of failure of an expedition, but also as a symbol for the many traces which Western explorers left behind. The contrast between the unmeasurable cruelty of the icy landscapes and the paper is comparable to the contrast between the tiny ship wreck on Caspar David Friedrich's painting and the devastating ice floes, as "a shipwreck that becomes the symbol of a landscape of death" (Böhme 2020, 352, transl. B.S.). "The aesthetics of the beautiful, nature and history reconciling nature has ended and gives room for the harsh experience of the post-Kantian sublime, of terror and destruction. The beautiful has fallen out of eternity. This is where modern art comes in" (Böhme 2020, 352, transl. B.S.). Today Polar ice is again connected to existential and destructive fate, but this time the terror comes from Western lifestyles and their emissions, caused by industrialisation.

It seems to me that no artwork can successfully fill the void of imagining the losses caused by global warming. The grand narratives of those days, for which the cookie wrapper stands, and which presented a man as a hero in the face of eternal ice, have been told out. The adventure associated with this story is the exploitation of humans, non-humans and fossil resources, an exploitation that became the cause of climate change. Therefore today it is revealed that the gaze into eternal ice never has been innocent, because we, the industrialised societies who are involved with colonial actions until today, are the hand that pushes us over into the abyss. It is important to respond to this process and art can help to feel and experience the responsibility. But it also becomes clear, that the changes are so big and vast, that feelings will always fail. Because it is impossible to respond in an 'appropriate' way, and this is, what German philosopher Günther Anders, in replacing the Kantian sublime in times of vast technological impacts, termed the "subliminal" ("überschwellig") (Anders 1980, 263). If art or expectations towards art rest in the mode of a sublime aesthetic, we won't be able to get over the heroic dimension of our own actions. But art can help to think carefully about which stories, which "cryonarratives" (Leane 2022) are told with and about the ice and which perspectives are addressed with the stories (un)told. This critical care as well applies to art, because art is also part of the aesthetics which becomes more and more fragile in times of global warming.

Bibliography

Anders, Günther. 1980. *Die Antiquiertheit des Menschen 1. Über die Seele im Zeitalter der zweiten industriellen Revolution*. Munich: Beck.

Antarctic Pavilion. 2015. About. http://www.antarcticpavilion.com/about.html. Accessed 26 Nov 2022.

Blumenberg, Hans. 1981. *Die Genesis der kopernikanischen Welt*. Frankfurt am Main: Suhrkamp.

Boetzkes, Amanda. 2020. Climate Aesthetics in the Ablation Zone. *Afterimage* 47 (2): 35–39. https://doi.org/10.1525/aft.2020.472007.

Böhme, Gernot. 1995. *Atmosphäre. Essays zur neuen Ästhetik*. Frankfurt am Main: Suhrkamp.

Böhme, Hartmut. 2016. Forces and Forms in Geo-Aesthetics. In *The Forces behind the Forms. Geology, Matter, Process in Contemporary Art*, ed. Beate Ermacora, Helen Hirsch, and Magdalena Holzhey, 37–50. Cologne: Snoeck Verlagsgesellschaft mbH.

———. 2020. Der Niedergang naturästhetischer Evidenz, oder: Über die Unvermeidlichkeit der Natur in den Künsten. *Kunstchronik* 73 (7): 340–357.

Brown, Andrew. 2014. *Art & Ecology Now*. London: Thames & Hudson.

Buckland, David. 2006. Cape Farewell Art and Climate Change. In *Burning Ice. Art & Climate Change*, ed. David Buckland, Ali MacGilp, and Sion Parkinson, 5–7. London: Cape Farewell.

Charrière, Julian. 2023. The Blue Fossil Entropic Stories. http://julian-charriere.net/projects/the-blue-fossil-entropic-stories. Accessed 10 Nov 2022.

Conference "Ice(St)Ages". 2021. https://www.uni-greifswald.de/en/university/information/annual-events/n/ice-stages-1-moving-environments-fleeting-encounters-and-performative-gestures/

DuMont. 1972. *Hans Haacke. Werkmonographie*. Cologne: Verlag DuMont Schauberg.

Fry, Edward F. 1972. Hans Haacke—Realzeitsysteme. In DuMont. 1972. *Hans Haacke. Werkmonographie*. Cologne: Verlag DuMont Schauberg.

Hannah, Dehlia. 2017. Hopeless Utopia. Enchantment and Contradiction in the First Antarctic Biennale. *Frieze* 188. https://www.frieze.com/article/hopeless-utopia

Holzhey, Magdalena. 2016. The Forces Behind the Forms. On the Exhibition. In *The Forces Behind the Forms. Geology, Matter, Process in Contemporary Art*, ed. Beate Ermacora, Helen Hirsch, and Magdalena Holzhey, 121–127. Cologne: Snoeck Verlagsgesellschaft mbH.

Hosseini, Anita. 2019. Olafur Eliasson's 'Ice Watch'. Bilderfahrzeuge. https://bilderfahrzeuge.hypotheses.org/3528

Kant, Immanuel. 1987. *Critique of Judgement*. Indianapolis: Hackett Publishing Co.

Leane, Elizabeth. 2012. *Antarctica in Fiction: Imaginative Narratives of the Far South.* Cambridge: Cambridge University Press.

———. 2022. Cryonarratives for Warming Times. Icebergs as Planetary Travellers in Ice Humanities. In *Ice Humanities. Living, Working, and Thinking in a Melting World*, ed. Klaus Dodds and Sverker Sörlin, 250–265. Manchester: Manchester University Press. https://doi.org/10.7765/9781526157782.00022.

Loomis, Chauncey C. 1977. The Arctic Sublime. In *Nature and the Victorian Imagination*, ed. U.C. Knoeplmaer and G.B. Temyson, 95–112. London: University of California Press.

McLuhan, Marshall. 1964. *Understanding Media. The Extension of Man.* New York: McGraw-Hill.

———. 1969. *Counterblast.* London: Rapp & Whiting.

Potter, Russell A. 2007. *Arctic Spectacles: The Frozen North in Visual Culture, 1818–1875.* Seattle: University of Washington Press.

Samman, Nadim. 2015. *The Antarctic Pavilion.* http://www.antarcticpavilion.com/about.html. Accessed 26 Nov 2022.

Silversea. 2022. https://www.silversea.com/. Accessed 10 Nov 2022.

Stifter, Adalbert. 1937–1938. Winterbrief aus Kirchschlag. *Am häuslichen Herd. Schweizerische illustrierte Monatsschrift* 41: 227–230.

Vienat, Bernard. 2018. Experience and Participation: Mathias Kessler's Conversational Impulse. In *Mathias Kessler. Staging Nature*, ed. Thorsten Sadowsky. Davos: Kirchner Museum.

Watt-Cloutier, Sheila. 2015. *The Right to Be Cold: One Woman's Story of Protecting Her Culture, the Arctic and the Whole Planet.* Toronto: Allen Lane.

Wichmann, Natalie. 2021. *Freezing Art.* https://www.schirn.de/magazin/kontext/2021/magnetic_north/freezing_art/. Accessed 26 Nov 2022.

Sensing Polar Ice Bodies

Stephanie von Spreter

She is not alone: the ice, too, is animate. (Singh Soin 2020b, 2)

From a larger-than-human-sized ice block, a ghost-like young woman becomes visible: her body and hair is wrapped in Space Age silver foil or, maybe from a more existential perspective, in the material that is used for emergency blankets.[1] Human and glacial body merge; the metallic body visually absorbs the glacial and vice versa. The ice block lies in an open gravelled landscape, with rough mountains in the back, as if it had been deserted following an invasion of some sort. In the distance, several other similarly shaped blocks are scattered around. All of these appear to have floated in the past, which is however no longer possible because the water

[1] In a conversation between the author and the artist, this double-connotation has been confirmed. Soin has additionally expressed that the material reflects the landscape back, and metaphorically also the viewer.

S. von Spreter (✉)
University of Tromsø, Tromsø, Norway
e-mail: stephanie.von.spreter@uit.no

A. Hemkendreis, A.-S. Jürgens (eds.), *Communicating Ice through Popular Art and Aesthetics*, Palgrave Studies in Media and Environmental Communication,
https://doi.org/10.1007/978-3-031-39787-5_7

has retreated. As a result, the ice blocks make up a loose assembly, convening around what remains: a large water puddle or a small stream, revealing smaller rocks from deep geological time (Fig. 1).

The described scene appears in a video that is part of the ongoing work *we are opposite like that* (2017–2022) by Himali Singh Soin (born 1987, based between London and Delhi). Singh Soin's five-year project springs from her participation in The Arctic Circle Residency programme to the Arctic archipelago of Svalbard aboard the barquentine tall ship Antigua in October 2017,[2] as well as a preceding expedition to Antarctica organised by Ibex Expeditions, India (February 2017). The project entails three videos (*we are opposite like that*, 2018; *we are opposite like that*, 2019; *How to Startle the Unbelieving*, 2021); two series of prints related to the first two videos (*we are opposite like that*, 2019; *Inverted Map*, 2020); an artist book

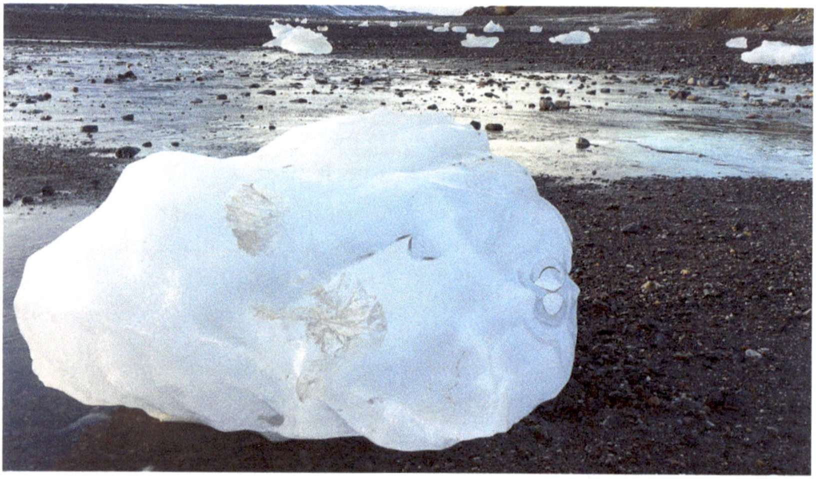

Fig. 1 Himali Singh Soin, *we are opposite like that*, video still, 2019 © Himali Singh Soin

[2] Svalbard was formerly called Spitsbergen. In 1920 Norway was granted sovereignty over the archipelago which resulted in a name change from Spitsbergen to Svalbard. Official maps name only the main island of the archipelago Spitsbergen today. However, Spitsbergen is still often used by (non-Norwegian) lay people for the entire archipelago. In addition, there is Russian disagreement about the naming and Russian officials prefer to use the old name.

(*we are opposite like that*, 2020), a text score print (*Boatness*, 2020); a sound work in three chapters, also published as a vinyl (*Subcontinentment*; *Lady Antigua*; *Antarctica was a queer rave before it got busted by colonial white farts*; all 2020); a tapestry visualising the sound wave of ice crystals smashing into one another (*Mountain, Pixelated in the Water*, 2021); an opera dedicated to the bird Arctic Tern (*an omniscience: an atmos-etheric, transnational, interplanetary cosmist bird opera spanning seven continents and the many verses*, 2022); a marble sculpture of Deception Island and a mandala performance (*Too Much and Not Enough*, 2022) as well as a series of live performances. Thus, *we are opposite like that* takes on a multiplicity of forms and materials, including poetry, sound, video, (still) photography, textile, sculpture, performance and prints.

According to the artist herself, *we are opposite like that* is an "ongoing series of interdisciplinary works that comprises mythologies for the poles, told from the non-human perspective of an elder that has witnessed deep time: the ice. It beckons the ghosts hidden in landscapes and turns them into echoes, listening in on the resonances of potential futures" (Singh Soin). Through this active statement, Singh Soin naturally pre-empts a reading of her work from a non-anthropocentric perspective, while her own self-identified "brown body" (Singh Soin) remains connected with the environment; on the one hand, this environment confirms her alienness, and on the other, her entanglement with it—she is inside the ice and the ice is inside her. As Åsberg and Braidotti compellingly express from a posthuman feminist perspective:

> In this new planetary age of the Anthropocene, defined by human-induced climatic, biological, and even geological transformations, we humans are fully in nature. And nature is fully in us. This was, of course, always the case, but it is more conspicuously so now than ever before: people are entangled in co-constitutive relationships with nature and the environment, with other animals and organisms, with medicine and technology, with science and epistemic politics. We live and die, play, thrive, and suffer by each other. (Åsberg and Braidotti 2018, 1)

This chapter aims to investigate how Himali Singh Soin's work *we are opposite like that* engages with posthuman feminist concepts within an Arctic discourse and in particular in relation to the climatic changes that also lead to the melting of the Polar caps. Its focus is on the gradual disappearance and cultural-ecological transformations of what has long dominated its landscape and mythologies: the ice. For this investigation, it is important to note that Singh Soin's "mythologies", as she describes them, are not on or about the poles—akin to the history of Western Polar imaginaries—but 'for' the poles (Singh Soin). Her artist book is explicitly dedicated *To Ice*, subtitled *We Tell your Story* (Singh Soin 2020b). Thus, the work becomes a caring dedication to the earth's seismic areas that are essential indicators for the well-being of our planet. In actively taking the perspective of the very material that (still) surrounds the poles, Singh Soin opposes the idea of the Arctic/Antarctic as lifeless, agentless matter. Here ice becomes both agent and metaphor for bodies in flux, in different aggregate states, both human and more-than-human.

In my process of disentangling and re-entangling the themes and concepts, my method uses still images and corresponding poetic quotes/voiceovers from Himali Singh Soin's video (2019)—the latter reappearing in the poem *how she became ice* published in her artist book (Singh Soin 2020a)—as intermissions and guiding storytelling elements. This choice is based on my conviction that the video and related book are central within Singh Soin's larger project. Here Singh Soin's poetry and fictive writing, film and photographic material and use of sound and historical illustrations are carefully woven together. Thus, we are provided with the multi-layered, multi-disciplinary stories, images and historical imaginaries of an Arctic for which Singh Soin advocates in her own description of the project. These quotes/voiceovers and images guide the reader through the work's interconnected themes and concepts from a posthuman feminist perspective, in the spirit of what Åsberg and Braidotti advocate: "Now, the tasks of the more-than-human humanities scholar are then to provide guiding stories with which to tell these stories, and to present adequate maps to the specifically situated historical locations" (2018, 5). These guiding storytelling elements are expanded by conversations between the artist and myself, which started with a 'love letter' I sent to Himali Singh Soin at the beginning of 2022.[3] This initial point of contact led to further conversations via email and recorded WhatsApp messages over the course of several months.[4]

[3] If one would like to contact the artist via her webpage the contact form is headed by the sentence "send me a love letter". See https://www.himalisinghsoin.com/contact (accessed 22 March 2022)

[4] If a below quote is uncited it emerged during the course of Soin and my informal communication between March and August 2022.

My chosen guiding elements map out the different themes and embodied concepts—or "figurations" as Neimanis calls them (2017b, 5)—that I identify in Singh Soin's project. These include the disappearance of planetary history through the melting Polar ice, and with it, the disappearance of the ice as a natural archive (Frank and Jakobsen 2019); Astrida Neimanis's figuration of bodies of water that I relate to bodies of ice (2017b, 5); the ghosts and monsters that haunt (Arctic) environments and are interlocutors for our present and future (Tsing et al. 2017b); the omnipresence of colonialism in the Arctic, including a type of delayed colonialism or slow violence (Nixon 2011) that affects the environment and its human and more-than-human inhabitants, more often than not carried north by a weather-and-water world of planetary circulation (Neimanis 2017a, 36); how the relics of historical Arctic exploration still haunt us today and how our situatedness points to our differences and distances from one another, but can also be used as a common feminist and transformative ground for creating other possible worlds and mythologies (Braidotti 2022, 3, 8). Thus, Singh Soin's mythologies sensitise the viewer/listener/reader to think with and learn from transforming Polar ice bodies—in relation to Neimanis's advocation to do this with water (Neimanis 2017a, 22)—to thereby equally transform ways of perceiving, feeling, knowing, understanding and re-imagining Polar landscapes and the unjustness felt in anthropogenically damaged environments.

> For historians the present had lost itself over time. (Singh Soin 2019, 00:22–00:26)

There is one image in Singh Soin's video that frequently re-occurs, though each time juxtaposed with different voiceovers: a rugged Arctic glacier reflected in surrounding water, seen as upside down to the viewer (Fig. 2). Because of its repetitive use, the image acts like a refrain in a poem, which not only creates rhythmic structure and familiarity, but makes it assume a central position within the video.

As a reflection in the water, the image of the glacier is blurred, making its surface not only appear carved by nature's forces but also reminiscent of sacred marble structures chiselled by humans. The glacier's colour hues—greys, beiges, light turquoises—enhance these marble-like qualities and evoke religious architecture from Singh Soin's native India, such as a

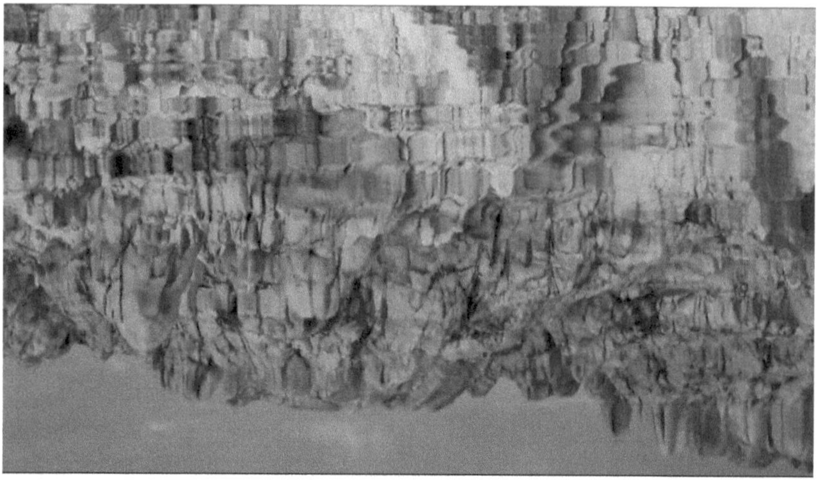

Fig. 2 Himali Singh Soin, *we are opposite like that*, video still, 2019 © Himali Singh Soin

Jain temple located in Ranakpur, Rajasthan. Temples are architectural structures that are simultaneously worldly and spiritual, material and immaterial, manifestations of belief systems and cosmologies and in movement towards spiritual worlds. Highlighting the carved, marble-like qualities of the glacier thus does not appear accidental. With Singh Soin's decision to film it as a reflection in the water—where boundaries between ice, water and sky blur—the image becomes a metaphor for the material and immaterial, of structures that are made and unmade by humans. Structures that we aim to control but ultimately get out of hand. Possibly that is why Singh Soin also included her own filmed footage from a former marble mining settlement on Svalbard later on in the video. Though abandoned, it is a popular place to visit. It tells the wondrous story of British-born venturer Ernest Manfield who, in the name of his co-founded Northern Exploration company, set foot on Svalbard in 1904 to search for precious metals and minerals. Following his claim to have found marble deposits, he set up a mine and called the settlement Ny London (New London). His endeavour, however, proved unsuccessful: once the marble was shipped and entered a warmer climate, it crumbled and turned to nothing but dust (Arlov 1989). Nevertheless, the mine operated (without profit) until 1920. Today, remnants of the settlement including some huts and disintegrating industrial tools speak of Svalbard's long history of extraction and exploitation of natural resources, as well as its human entanglements.

The blurred and inverted image of the glacier with its marble-like qualities then speaks about its past, its entanglement with human bodies and its gradual disappearance. It attends to the material interconnections between the human and more-than-human world, in what Alaimo would describe as trans-corporeal *bodily natures* (Alaimo 2010). While the glacier itself is a body that transforms into and enters other bodies through the process of melting, the pixelated image in the video metaphorically visualises this process. According to Singh Soin, this visualisation became intentional: "but I was really fascinated by it looking like pixels and [...] by this feeling that it looks like the very technologies that have created it are in fact creating its death; that is to say climate change is essentially created out of the pixelation of the environment". In this sense, the pixelated view points to the need to acknowledge our responsibility for the unmaking of our environment.

This 'glacier death' has also consequences for history-writing. Because of irreparable environmental damage, we find ourselves at a historical threshold. Singh Soin expresses in the voiceover to the pixelated glacier image: "For historians the present had lost itself over time", while the rhythmic sound of plucked cello strings evokes a slow but continuous dripping of melting ice (Singh Soin 2019). What does it mean when the present had lost itself over time? If the present has lost itself, is a future even thinkable? How can historians then grab a present that is soon to be history, and a present that is supposed to shape the future? It appears that the present, and with it, its past, slips through historians' fingers. It melts away before their very eyes.

In their introduction to *Arctic Archives: Ice, Memory and Entropy*, Frank and Jakobsen consider the Polar regions to be the knowledge archive of our planet (2019). Here ice acts as a memory medium to form a natural archive that provides us with information on climate history and deep time. However, when it melts, it wipes out the histories it formerly stored. Consequently, Frank and Jakobsen argue, climate change not only threatens our future but also evidence to what lies buried from the past. Thus, we are standing at this threshold, or tipping point, where both our past and future are about to collapse. Once the ice melts during this time of "Great Acceleration", it will no longer be possible to retrieve those hidden histories in its material form (Frank and Jakobsen 2019, 16). Thus, historians are met with a dilemma. Their primary source material disappears.

Ice and permafrost, Frank and Jakobsen explain, differ from other natural archives in terms of which ages they conserve and in what form that happens. While stone receives only very old layers, the ice preserves younger layers of geological and historical facts. It allows for an extremely precise dating and protectively preserves atmospheric conditions and life-forms of the past. However, it is the most vulnerable and unstable natural memory medium because it depends on sub-zero temperatures (Frank and Jakobsen 2019, 9–10). Thus, when bodies of ice melt also the primary historical evidence 'dies'. Singh Soin's filmed glacier then sensitises the viewer for standing at this threshold where temporalities collapse and distinctions between the human and more-than-human blur; the glacier is a ghostly, disappearing icy body where nature and culture merge. Singh Soin's video thus not only communicates how this icy body makes nature and technology collapse, but also how historical evidence disappears and thereby we ourselves threaten to become a barely traceable history.

> The ice moved through her for a few hours. A mineral messenger. (Singh Soin 2019, 05:25)

In the meantime, the silver-wrapped Polar ice body has moved on, the ice block having become a shadow of itself. "The ice has moved through her for a few hours", the voiceover tells us (Singh Soin 2019). And indeed, the silvery brown body begins to wander through the darkening Arctic landscape, no longer encapsulated in the ice. Now the silver foil appears like an enveloping protective layer, like the emergency blanket meant to keep the body warm—warm enough for the ice to become fluid and move through her as a "mineral messenger" (Singh Soin 2019).

In *Bodies of Water: Posthuman Feminist Phenomenology*, Astrida Neimanis advocates rethinking and re-imagining embodiment from the perspective of our bodies' wet constitution, inseparable from the planetary crisis we find ourselves in. Consequently, we have to give up the idea of our bodies as autonomous, as they are always also—materially and conceptually—more-than-human (Neimanis 2017b, 1–2). Water moves through our bodies and determines our specific situatedness. Water contains information, it contains life-giving and (human-made) toxic substance(s) that enter and leave our bodies to again enter and leave other (more-than) human bodies. In its frozen state, water is considered a mineral; thus this mineral messenger carries stories, substances and data from one body to another. We are situated bodies of water.

> Ice, a shadowy interlocutor of what has been, what is, and what will come, is the phantasmagoria of an alien otherworld. (Singh Soin 2019, 07:41)

The Arctic has been subject of myths and mythologies since human existence; an imaginary place that for those who lived far away or tried to reach it—such as for the ancient Greeks—was "[...] a fabled land where myth and geographical theory fused" (McGhee 2007, 22). For them the North Pole and the Pole Star always earned an important place in geography and mythology. In Ptolemy's cosmography, for example, the poles were considered oppositional fixpoints for holding the earth together on its central axis within the cosmos. The poles were the mythical origin of time and space. Even Polar explorer Robert Peary, who claimed to be the first to reach the North Pole in 1909, considered the North Pole to have mythical powers, despite firm beliefs in technological and scientific progress. According to Michael Bravo, Peary saw the giant Antaeus—son of the sea god Poseidon and earth goddess Gaia—as a symbol for this mythical power emerging out of the Polar ice (2019, 191, 188). Singh Soin juxtaposes this image of the rising Antaeus with historical aurora sketches (*Observations faites au Cap Thordsen, Spitsbergen, par l'expédition Suédoise*, 1882–1883) and her own film footage of Svalbard's topography in her video (Fig. 3). In the historical image, Antaeus appears as the creator and guardian of the land. Both the ice and the aurora lights frame his figure, placing him centre stage. This central position is further enhanced by his headgear, a laurel crown and an icicle pointing towards the sky. His young face and figure are veiled, heightening the perception that he is a mythical figure with great powers, ruling over the North Pole and its surrounding land of ice.

Other ancient texts also considered the poles' mythological and life-giving powers. In the Hindu Puranas, a sacred mountain is imagined at the centre of Hyperborea (Mount Meru). The mountain rests on four pillars made of gold, iron, silver and brass, each of which points to the cardinal points of the compass. High up in the sky above, at the feet of Vishnu near the Pole Star, the River Ganges originates. The celestial and

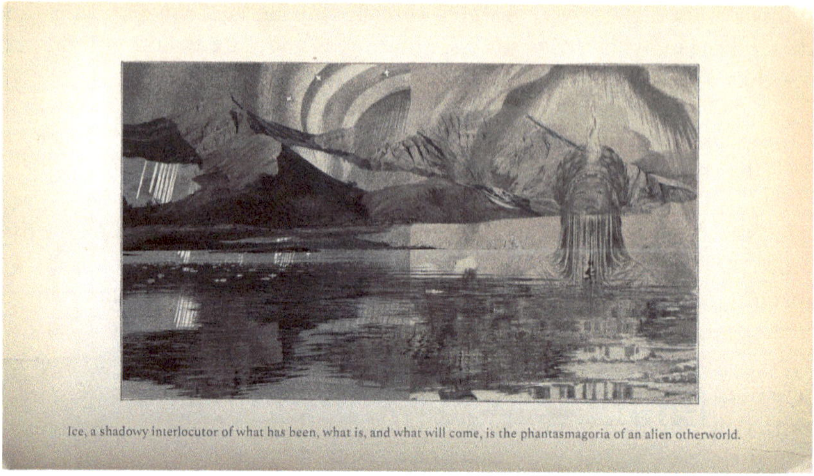

Fig. 3 Himali Singh Soin, *we are opposite like that*, video still, 2019, © Himali Singh Soin

terrestrial area around the North Pole becomes the physical location of the source of life (Bravo 2019, 138).

Within local indigenous mythologies, however, the North Pole and the Pole Star play a minor role. With shifting horizons at high latitudes, the fixed Pole Star and the geographical North Pole are not useful for navigating a system of trails on the ice. Rather, the surrounding, moving stars—in dialogue with the icy landscape/horizon line—play an important role for navigation. Here myths and legends are crucial for survival, naturally and culturally:

> Sometimes, the story of a trail will involve or take the form of a myth or legend that may give meaning and shape to a trail. Similarly, the trail of a constellation that can be tracked across the night sky is described in a myth that explains this movement. However, rather than thinking of these myths as a stationary form of mapping, one wants to keep in mind that Inuit travellers are using a moving frame of reference that changes as they themselves move. If one thinks of the Inuit world in terms of the fluid movement of people, animals and spirits across intersecting or connected systems of trails, one begins to gain a better sense of how the stars figure in their navigation tradition. (Bravo 2019, 20–21)

Thus, the icy myths that might appear to outsiders as the phantasmagorias of an alien otherworld are an integral element of survival for northern indigenous populations. The ice—the shadowy interlocutor—is in dialogue with the elements in the sky, serving as a cultural and historical messenger and guarantor of the continuity of life. However, with climate change, also these myths and tools of survival are threatened. Recent interviews with Inuit hunters and elders reveal that:

> over the course of their lifetimes it appears that the world has tilted on its axis. Accustomed to watching the sky for weather patterns amid diurnal seasonal changes, the Inuit have noted longer periods of daylight, even as much as a full hour longer, in the short days of winter. [...] And while the sun rises from the same location it always did during the calendar year, it appears to set at a different location on the horizon than it did a generation ago. [...] the sun seems to be higher and hits more directly since the tilt. Hunters are especially conscious of the daylight hours and the quality of light because of the narrow margins of opportunity to hunt and fish in the winter. So while it is possible to measure global warming through ice core analysis and sea temperature changes, the fundamental changes of orientation for the Inuit are taking place in the sky and atmosphere, the site at which they "read" and perceive the landscape. (Boetzkes 2018, 135)

In Singh Soin's video the ice becomes an interlocutor between divergent worldviews and mythologies. It is the protagonist and the 'elder' that has served, witnessed and provided both indigenous and external cultures in the Arctic with "what has been, what is, and what will come" (Singh Soin 2019). It is metaphor, mediator and material at the same time. And, as Singh Soin formally demonstrates in the video, it is where contemporary and historical material, Arctic imaginaries and Singh Soin's own filmed, overlaid footage merge into one another. These are also the images of an Arctic environment which show how colonisation and anthropocentric worldviews are 'ever present' ghosts of the past and in fact contribute to creating the monsters of our present.

In their introduction to *Arts of Living on a Damaged Planet* Bubandt, Gan, Swanson and Tsing write: "Our monsters and ghosts help us notice landscapes of entanglement, bodies with other bodies, time with other times. They aid us in our call for a particular approach to noticing—one that draws inspiration from scientific observation alongside ethnography and critical theory" (2017a, M7). In this endeavour, they advocate slowing down and listening to the world, empirically and imaginatively at the same time. They encourage us to think that ghosts and monsters can help us find our way when there is environmental damage, in fact trace them. In Singh Soin's words, we must realise that ice is a phantasmagoria of an alien otherworld. But the truth is that we are already inside this alien otherworld in which we are and create the monsters. And the melting bodies of ice become their shadowy ghosts.

> The viscous material of an encroaching ice sheet is haunted by colonialism calving. (Singh Soin 2019, 08:44)

If the alien otherworld is a world that is not so alien after all, but that we are right inside it—which we can see through sensing its ghosts and monsters—we come to understand how the "viscous material of an encroaching ice sheet is haunted by colonialism calving" (Singh Soin 2019). In employing the word calving, Singh Soin references the terminology used when glaciers lose part of their mass. In this process, a large chunk of ice at a glacier's edge crashes into the water, resulting in a booming sound and large waves. Like a cow's calf, the resulting iceberg then moves into the world. The 'ice calf' floats, melts and merges with its environment, with other bodies. In Singh Soin's work, however, it is not the glacier calving, but colonialism, chunks of which are carried further into the world. And it is indeed the ice sheet, and Arctic, that is haunted by these colonial and imperialist chunks, still today. As Neimanis has expressed, colonialism can be remote: "Indeed, the Anthropocene may also index an important mutation in forms of colonial power, where colonisers need not physically occupy a place with their discrete bodies for the environmental effects of (neo)colonial power to be felt" (2017a, 163). Neimanis uses several examples of how life and the environment in the Arctic are affected by this coloniality "at a distance" (2017a, 165), such as the fact that permafrost degradation raises the mercury levels in the Arctic food chain. Another scientifically proven example is the bioaccumulation of various anthropogenic contaminants that make their way north from warmer regions, which then become entrenched in the fat of Arctic sea mammals. As a result, they enter the food chain of Inuit women, who, in turn, transfer the toxic waste to their children through their breastmilk. Neimanis calls this effect "body burden" (Neimanis 2017a, 36, 164) which is socially, culturally and gender determined. She further underlines:

> While the Arctic is "generally considered to be one of the last pristine regions on Earth", its populations (human and more-than-human) again bear the brunt of global human imperialism. This kind of incursion 'at a distance', precipitated by massive fossil fuel burning, consumption, and toxic release into planetary waters and weathers (out of sight, out of mind), traces new vectors of coloniality, and emerging markers of vulnerability and survivance (Vizenor 1999) across difference. (2017a, 165)

Equally Tsing et al. note this Anthropocentric presence of the ghosts in our environment and use the example of human-made radiocesium that travels in water and soil, and gets inside plants and animals (Tsing et al. 2017b, G2). We cannot see it, although we can trace it throughout ruined landscapes.[5] This effect is comparable to Neimanis's coloniality "at a distance" (2017a, 165), which in turn is related to the process of 'slow violence', a term coined and defined by Robert Nixon as a violence that "occurs gradually and out sight, a violence of delayed destruction that is dispersed across time and space, an attritional violence that is typically not viewed as violence at all" (2011, 2).

In Soin's video, the sentence "The viscous material of an encroaching ice sheet is haunted by colonialism calving" appears as a caption underneath a wood engraving by Gustave Doré, which Singh Soin appropriated from the book *London: A Pilgrimage* (1873) (Fig. 4). The engraving shows a man sitting on a boulder at the uncultivated shore of the Thames with a drawing board and pen in hand, overlooking a London in ruins on the other side. Depicted in profile, the viewer can see that the man has a goatee beard and is dressed in an oriental cloak and head cloth—none of which were fashion amongst English noblemen at the time. Clearly, this man is not a Western body from the Victorian civilisation that is now found in ruins. Rather, this is a man who comes from the colonial periphery to its centre. As a colonised body he brings the 'calving' effects of colonialism back to its instigator. Doré's engraving is entitled *The New Zealander*, and there are different interpretations as to why it bears this

[5] Michael Bravo uses further examples of slow violence/ghostly environmental damage in the Arctic: "The Barents Sea off the coast of Norway and Russia is reportedly the most radioactive in the world, largely as a result of nuclear atmospheric tests carried out during the Cold War, emissions from reprocessing plants and the accident at Chernobyl. To the north and east of the Greenland and Barents Seas, high concentrations of old plastic arrive from the Atlantic Ocean by thermohaline ocean circulation, which acts as a 'plastic conveyor belt' from distant sources" (2019, 212).

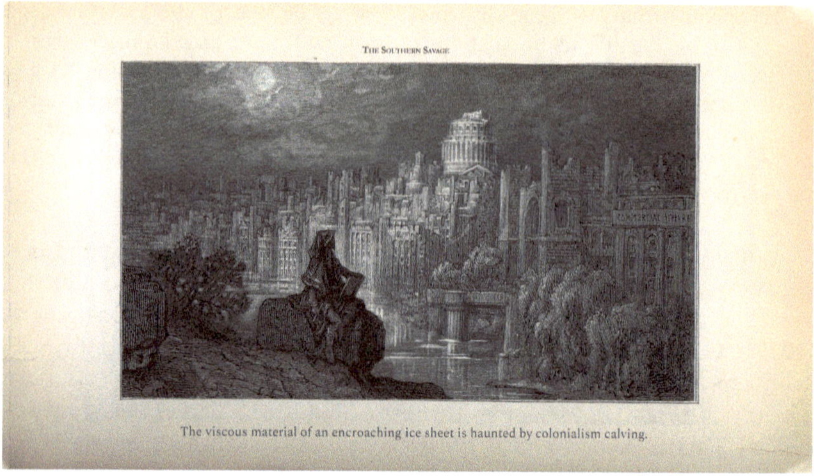

Fig. 4 Himali Singh Soin, *we are opposite like that*, video still, 2019 © Himali Singh Soin

title, while all point to the biography of a person coming from the British colony (Arnold 2017; Dobraszczyk 2017; Skilton 2014). In the video Singh Soin replaces this title with another—*The Southern Savage*—and thus takes a more overt and critical stance as to how the colonised body is perceived from a Eurocentric perspective. At the same time, this body becomes witness to the ruins of Western civilisation, taking a chunk of colonialism and its destructive forces back to its origins. Singh Soin expresses in our conversation: "In the Global North I suddenly become a brown body that is representative of many brown bodies and the loss that they have encountered". So how should one act as a colonised body in the ruined landscape of the Arctic with which Singh Soin technically animates Doré's engraving? The Southern Savage has pen and paper in hand. Maybe he imagines another future, worlds another world? Singh Soin expresses: "There seems to be this sense that, especially when you come from a place like the Global South, where you just witness always just so much atrocity and where every life is not valued in the same way it is in the Global North, then all you are left with is this kind of dream for joy or something".

> Her broken chronometer preserved in ice, still recording two types of error: the imperfection of the image exposed / and the great distance it represented. / When examined with a telescope it proved to be our distance from one another. (Singh Soin 2019, 09:18–09:32)

As much as colonialism haunts the Arctic and colonised bodies, there are objects coupled with it. The marine chronometer might appear to be an insignificant scientific object at first, yet it is not. Only when this precision timepiece was invented in the early eighteenth century, did it become possible to determine a ship's position on the meridian, which happened by calculating the difference between Greenwich Mean Time (GMT) and the ship's local time (which in turn could be calculated with the help of a sextant). Together with the ship's latitudinal position—in the Northern hemisphere determined with the help of the North Star (Polaris), a compass and a quadrant—a ship's precise coordinates could be verified. This made it also possible to determine further destinations. Thus, the chronometer proved to be an essential Western invention to enable the Age of Discovery and imperialist expansion via the oceans. It thereby also—if not exclusively—contributed to the acceleration of colonialism, resource extraction and subsequent (environmental) damage.

On 1 July 2002 the Norwegian government enacted a law that all remnants of human activity on Svalbard prior to 1946 are classified as cultural heritage. These must not be removed from its premises and are automatically the property of the Norwegian state. The law, Svalbardmiljøloven (Svalbard Environmental Protection Act), applies to both fixed elements (such as trapper's cabins, coal mining rails etc.) and loose objects including those that emerge arbitrarily or via excavation (Lovdata 2022). With the melting of the ice, loose objects from historical Polar expeditions formerly preserved in the ice or underneath in the ocean now also emerge.[6] The 'Arctic archive' releases these objects in almost the same condition as

[6] In 2014 and 2016 for example, the ships HMS Erebus and HMS Terror were found in the Arctic Ocean, having been lost for over 160 years. Used under the command of Sir John Franklin in search of the Northwest Passage, it is documented that they were abandoned in 1848 and the crew was never seen again. This lost expedition triggered the largest search and rescue endeavour seen in Arctic history, with 32 directly motivated expeditions embarking between 1847 and 1859 (Ross, 57). With the retreating ice it was finally possible to find the two vessels with archaeologists finding many objects on board that are almost intact. Many of these objects are today found at the Royal Museums Greenwich.

when they were in use. Since the chronometer was one of the most important scientific instruments on board expedition ships, and often having several on board in fear of becoming dysfunctional, a broken chronometer—as Soin imagines it in her poetic voiceover—can well appear from the ice.

Singh Soin's/the ice's broken chronometer, however, is not visible in her video. Instead, we are forced to imagine what the chronometer would look like and in fact wonder how a time-measuring instrument used for determining one's longitudinal position can expose an (imperfect) image. In Singh Soin's video, we witness the mountainous landscape of Svalbard passing by as if we were on a ship, overlaid with organic, almost transparent ice formations. The landscape appears as if it were from out of space, far away. Technically, the ice acts like a lens on top of the landscape, blurring and distorting it. Maybe this is what Singh Soin describes as "the imperfection of the image exposed" (Singh Soin 2019): an image that appears—because of the 'ice lens'—out of focus and which thereby puts an imaginary distance between the Arctic landscape and us. Or maybe the chronometer measures the temporal and geographical distance between the ice and our physical bodies, revealing how distant yet connected our realities are. The telescope, besides being another important scientific instrument for maritime navigation, observes this poetically described condition of distance.

The video's final scenes again show Singh Soin (as the elder) moving through the Arctic landscape. Still wrapped in a silver blanket and turban, we observe her eyes fixing a point in the distance. It appears as if she imagines another world to exist in, another reality (Fig. 5). Maybe here the ice becomes the carrier-bag in which—as Ursula K. Le Guin has expressed in her carrier-bag theory for science fiction—Singh Soin's imagined chronometer can tell the time and place of another world (1986). However, while Singh Soin imagines a futurist otherworld, it is clear that these spring from the very real politics of location we find ourselves in. As Braidotti makes aware, we differ in terms of locations, our access to environmental, social and legal entitlements, technology, safety, prosperity and good health services. She says that "we" are "'we'-are-not-one-and-the-same-but-we-differ"; and even more that: for posthuman feminists "we" means "'we'-who-are-not-one-and-the-same-but-are-in-this-together" (2022, 8). But in order to prove that these differences exist and instigate changes we

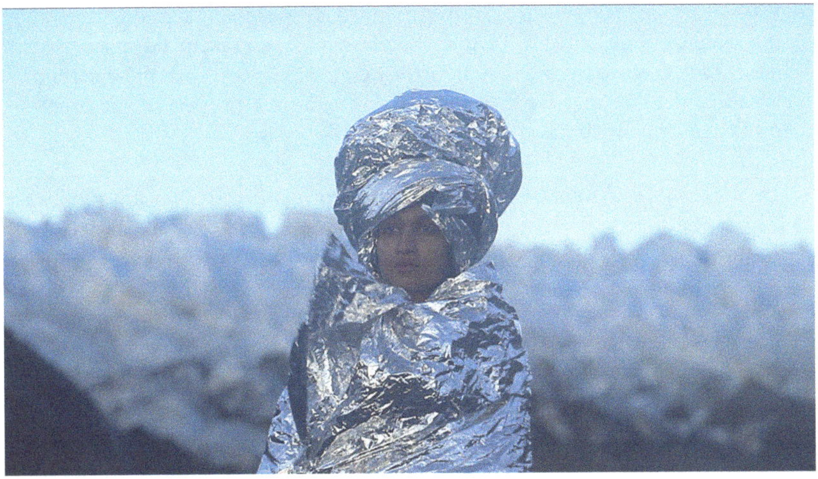

Fig. 5 Himali Singh Soin, *we are opposite like that*, video still, 2019 © Himali Singh Soin

need science and its instruments. Scientific research in(to) the ice not only discloses that climate change is human-induced, but also uncovers how its effects are unevenly distributed and often connected to race, class and gender, as I have shown is also the case in the Arctic. It can measure how toxins travel in and through bodies of water and are found in bodies of ice. It can, in fact, make visible the ghosts from the past. And it can make science connect to art and fiction in order to imagine, in Singh Soin's words, a less alien otherworld.

Epilogue

Soin's artistic practice aligns itself with futurisms, science fiction and speculative fabulation. In fact, Soin calls her South Asian futurism 'Subcontinentment', a word created from 'subcontinent' and 'contentment'. It 'aligns' itself with Afrofuturism, Sinofuturism, Ethnofuturism and Indigenous Futurism (Singh Soin 2020b, 17). For Singh Soin, Futurism is not related to

European Futurism which is so much linked to speed and acceleration and the violence that comes with that. But instead, is a kind of Futurism that looks back at the past and says: how do we move forward with these certain wisdoms that we may have accumulated. But also letting go of the prejudices that came with that timeline.

In *Walking the Clouds: An Anthology of Indigenous Science Fiction* Grace Dillon puts forward that in Indigenous Science Fiction literature the Apocalypse has already happened, and that it is in that sense post-Apocalyptic. She writes:

It might go without saying that all forms of Indigenous futurisms are narratives of 'biskaabiiyang', an Anishinaabemowin word connoting the process of "returning to ourselves", which involves discovering how personally one is affected by colonization, discarding the emotional and psychological baggage carried from its impact, and recovering ancestral traditions in order to adapt in our post-Native Apocalypse world. (Dillon 2012, 10)

Thus, this kind of storytelling shows ruptures, scars and trauma but also provides healing. It aims to return to a state of balance through the condition of resistance and survival (survivance, as coined by Gerald Vizenor). It does have the capacity to envision native futures, indigenous hopes and dreams recovered by rethinking the (colonial) past in a new framework (Dillon 2012, 2).

Similarly, the use of fiction and speculative fabulation in posthuman feminism and feminist new materialisms are used as tools for empowering the dispossessed and impoverished, a "decolonial and radical struggle to affirm positively the differences among marginalized people(s). [...] It means creating other possible worlds" (Braidotti 2022, 3). Tobias Skiveren underlines that these tools are particularly attractive to transcend anthropocentric regimes of truth:

Fictionalizing non/human entanglements, then, allows new materialist thinkers to, at least momentarily, sidestep questions of falsehood and truthfulness, while at the same time proposing new post-anthropocentric ontologies by imaginative and affective means. The stories adapted here are supposed to convince, not because they are true, but because they do not have to be. Readers are provided an opportunity to imaginatively and affectively sense a world in which the non-human is partly human, and the human is partly non-human, without having to believe that a spool of thread can run and talk in any literal sense. (Skiveren 2022, 5–6)

I advocate that Singh Soin's artistic practice aligns with both. As a protagonist and more-than-human storyteller, the Polar ice instigates affect and assumes the role of being a knowledgeable agent, material and immaterial at the same time. But of course, it is Singh Soin herself who embodies the ice through her own brown body. Singh Soin fabulates about a world elsewhere because she knows it does not exist in the here and now. But it can be imagined. She says: "it definitely feels like when you notice these cracks in the world, they can either be these fissures in which everything falls apart. Or they can be an opening through which the light is let in and they can be an opportunity to imagine." It is this imagination that can open the cracks and can communicate (scientific) knowledge about a colonialism that is not so distant after all, while it can also open up visions of future worlds beyond climate change.

Bibliography

Alaimo, Stacy. 2010. Bodily Natures. In *Bodily Natures: Science, Environment, and the Material Self*, 1–25. Bloomington & Indianapolis: Indiana University Press.

Arlov, Thor B. 1989. *A Short History of Svalbard. Vol. 4, Polarhåndbok (Printed Edition)*. Oslo: Norsk polarinstitutt.

Arnold, Dana. 2017. Misprisions of London. *Art History* 40 (4): 770–783. https://doi.org/10.1111/146-8365.12336.

Åsberg, Cecilia, and Rosi Braidotti. 2018. Feminist Posthumanities: An Introduction. In *A Feminist Companion to the Posthumanities*, ed. Cecilia Åsberg and Rosi Braidotti, 1–22. Cham: Springer International Publishing.

Boetzkes, Amanda. 2018. How to See a Glacier in a Climate Landscape. *Weber – The Contemporary West* 35 (1): 123–137.

Braidotti, Rosi. 2022. *Posthuman Feminism*. Cambridge: Polity Press.

Bravo, Michael. 2019. *North Pole: Nature and Culture*. London: Reaktion Books Limited.

Dillon, Grace L. 2012. *Walking the Clouds: An Anthology of Indigenous Science Fiction*, Sun Tracks. Vol. 69. Tuscon: University of Arizona Press.

Dobraszczyk, Paul. 2017. Sunken Cities: Climate Change, Urban Futures and the Imagination of Submergence. *International Journal of Urban and Regional Research* 41 (6): 868–887. https://doi.org/10.1111/1468-2427.12510.

Frank, Susi K., and Kjetil A. Jakobsen. 2019. Introduction. In *Arctic Archives. Ice, Memory and Entropy*, ed. Susi K. Frank and Kjetil A. Jakobsen, 9–17. Bielefeld: transcript Verlag.

Le Guin, Ursula K. 1986. *The Carrier Bag Theory of Fiction*. The Anarchist Library. https://theanarchistlibrary.org/library/ursula-k-le-guin-the-carrier-bag-theory-of-fiction. Accessed 14 July 2022.

Lovdata. 2022. Svalbardmiljøloven. https://lovdata.no/dokument/NL/lov/2001-06-15-79. Accessed 14 July 2022.

McGhee, Robert. 2007. *The Last Imaginary Place: A Human History of the Arctic World*. Chicago: University of Chicago Press.

Neimanis, Astrida. 2017a. *Bodies of Water: Posthuman Feminist Phenomenology*. London: Bloomsbury Academic.

———. 2017b. Introduction: Figuring Bodies of Water. In *Bodies of Water: Posthuman Feminist Phenomenology*, 1–26. London: Bloomsbury Academic.

Nixon, Rob. 2011. *Slow Violence and the Environmentalism of the Poor*. Cambridge, MA: Harvard University Press.

Ross, W. Gillies. 2002. The Type and Number of Expeditions in the Franklin Search 1847–1859. *Arctic* 55 (1): 57–69.

Singh Soin, Himali. 2019. *We Are Opposite Like That*. Video, 12:54 min

———. 2020a. How She Became Ice. *We Are Opposite Like That*, 36–37. Delhi: Subcontinentment Press.

———. 2020b. *We Are Opposite Like That*. Delhi: Subcontinentment Press.

Singh Soin, Himali. Website. https://www.himalisinghsoin.com/we-are-opposite-like-that. Accessed 06 Aug 2022.

Skilton, David. 2014. Gustave Doré's London/Londres: Empire and Post-imperial Ruin. *Word & Image* 30 (3): 225–237. https://doi.org/10.1080/02666286.2014.938528.

Skiveren, Tobias. 2022. Fictionality in New Materialism: (Re)Inventing Matter. *Theory, Culture & Society* 39 (3): 187–202. https://doi.org/10.1177/0263276420967408.

Tsing, Anna Lowenhaupt, Heather Anne Swanson, Elaine Gan, and Nils Bubandt. 2017a. Introduction. Bodies Tumbled into Bodies. In *Arts of Living on a Damaged Planet: Ghosts of the Anthropocene*, ed. Anna Lowenhaupt Tsing, Heather Anne Swanson, Elaine Gan, and Nils Bubandt, M1–M12. Minneapolis/London: University of Minnesota Press.

———. 2017b. Introduction. Haunted Landscapes of the Anthropocene. In *Arts of Living on a Damaged Planet: Ghosts of the Anthropocene*, ed. Anna Lowenhaupt Tsing, Heather Anne Swanson, Elaine Gan, and Nils Bubandt, G1–G14. Minneapolis/London: University of Minnesota Press.

Antarctic Science on the Musical Stage

Hanne E. F. Nielsen, Elizabeth Leane, Dana M. Bergstrom, and Carolyn Philpott

Antarctica is a remote, extreme continent visited by few. Nevertheless, it is a place alive in people's imaginations. Much like the ice that covers the continent, this imagined version is not static—rather, it responds to cultural values and priorities that dominate at different times. Since the signing of the Antarctic Treaty (1959), science has been the primary reason for human presence on the continent, with research increasingly revealing the interconnections between the Antarctic region and the rest of the world. For example, the continent's seasonal changes play a major role in driving global ocean currents, and the melting of southern ice sheets threatens low-lying areas much further north (DeConto et al. 2021). Anthropogenic change that impacts the far south and its ice therefore has global

H. E. F. Nielsen (✉) • E. Leane • C. Philpott
University of Tasmania, Hobart, TAS, Australia
e-mail: hanne.nielsen@utas.edu.au

D. M. Bergstrom
Australian Antarctic Division, Kingston, TAS, Australia

A. Hemkendreis, A.-S. Jürgens (eds.), *Communicating Ice through Popular Art and Aesthetics*, Palgrave Studies in Media and Environmental Communication,
https://doi.org/10.1007/978-3-031-39787-5_8

implications. These implications, however, and the science that helps us articulate these, are not always widely understood (see e.g. Priestley et al. 2021, 5).

In an analysis of art/science connections in Antarctica, Craig Stevens et al. argue that "[a]rt may be made as a guide to understanding sense of place, and also as a pathway to understanding and valuing scientific ideas" (2019, 289). While the term 'art' often evokes an idea of high culture, popular artistic works are arguably just as valuable—if not more valuable—in their ability to connect with wide audiences. Here we ask how a popular stage musical can make global science challenges accessible to a wide public and help build connections with the Antarctic for those who will never see it. We begin with a contextual analysis of the musical as a form of science communication and a short history of Antarctica's depiction on stage. We then turn to our example, *Antarctica—A New Musical* (2016), which was co-created by songwriter and composer Dugald McLaren and Antarctic scientist Dana Bergstrom, focusing particularly on its engagement with scientific characters and concepts. Drawing on drafts, early recordings and reviews, we trace the development of this musical from concept to its debut in the Theatre Royal in the Antarctic 'gateway' city of Hobart in Tasmania (Australia). Finally, we consider a recent podcast version of the work, *Antarctica, Beneath the Storm,* in which scientific questions have become increasingly central, while the stage itself has become redundant.

Tracing the evolution of the musical allows the work to be viewed as an iterative experiment in communicating science. While the musical was described by workshop director Allan Jeffries as being about "ordinary people in an extraordinary place" (opening remarks, workshop performance, Peacock Theatre Hobart, 2010), we suggest it is just as much about the place itself. As the editors of the recent collection *Performing Ice* suggest, "[i]cy environments [...] share with human performances a contingent and unpredictable quality, making them both challenging and productive to consider in the context of performance studies" (Leane et al. 2020, 2). In this sense, the iterations of *Antarctica—A New Musical* mirror the shifting, flowing, melting southern ice: "The icy poles are melting quick/could soon be falling off the stick" (McLaren, arranged by Wood 2016). This musical work speaks to both scientific and artistic questions around our relationship with the far south and with the more-than-human world. By demonstrating how the popular form of the musical can bring the far south into the lives of those 'back home', this chapter presents insights for environmental and scientific storytelling about remote places.

The 'Science Musical': A Neglected Genre

While scientific ideas and processes have inspired theatrical productions for centuries—Christopher Marlowe's *Dr. Faustus* (1604) is an early example—the last few decades have seen a particular focus on the 'science play'. In her book *Science on Stage* (2006), Kristin Shepherd-Barr argues that the "surge of new plays" catalysed by the success of Michael Frayn's *Copenhagen* (1998) is a "true phenomenon", merging the arts and sciences like no other genre (2006, 1). For Shepherd-Barr, science plays "share certain critical features: A casting of the scientists as hero or villain (or sometimes both), a direct engagement with 'real' scientific ideas, a complex ethical discussion, and an interdependence of form and content that often relies on performance to convey the science" (2006, 2). *Copenhagen*, a play about a wartime meeting of two of the key architects of quantum mechanics, is a particularly good example. Not only do the characters explicitly enact subatomic processes on stage, but their attempt to recreate a historical moment—played out several times both within the play and every time it is performed—explores questions inherent to quantum theory, such as uncertainty and the impact of the observer on the observed. For Shepherd-Barr, this integration of scientific ideas with the possibilities of theatrical performance is key to the success of recent plays.

Few musical works feature in analyses of scientific performance. Shepherd-Barr's bibliography lists over 120 science plays, but only about 10% are musicals or operas and her study largely ignores them. Implicit in her analysis is the seriousness of the 'science plays' that she examines, and the "sort of science and theatre event that combines music with science in a lighter spirit" goes largely unexamined (2006, 237). Another, more recent collection, Martin Willis's *Staging Science* (2016), similarly sidelines musical theatre, even while emphasising the diversity of science performance in the contemporary period, including science television and "science stand-up" (2016, 7).

This neglect of the 'science musical' (to use Shepherd-Barr's term) can be tied to a wider marginalisation of the musical within theatre and performance studies. Musical theatre began to attract sustained academic attention only in the very late twentieth century, and "the notion that musicals, including the music and lyrics, script and performance, deserve scholarly attention and place in academia is still not accepted in many universities" (Wolf et al. 2022, 202). This is despite musicals being a highly popular—and hence potentially influential—genre and often the "only theatre [many people] have experience[d] as participants or as spectators" (Wolf et al. 2022, 201). The very popularity of musical theatre means that not only is it considered non-serious as art, but also that commercial-in-confidence aspects of its production and performance make access to detailed information difficult (Wolf et al. 2022, 202). In addition, the combination of different art forms in musical theatre makes it "extraordinarily complex to study" (Wolf et al. 2022, 202). This is even more the case when musicals are also concerned with scientific methods and data. For these reasons, the present authorship team is not only interdisciplinary but also includes one of the creators of the musical *Antarctica*.

Reflecting contemporary social, environmental and political concerns, the 'climate change play' is a fast-emerging genre, to the extent that one critic writes, "[t]he world of drama and performance has been inundated with new works that address anthropogenic climate change" (Balestrini 2017, 114). This genre has produced a corresponding scholarship, often located within ecotheatre or ecodramaturgy, as well as the environmental humanities more generally. Pointing to Steven Waters's *The Contingency Plan* (2009) as a "watershed" text, Adeline Johns-Putra argues that climate change plays have two key characteristics: first, they usually depict or are set in the aftermath of a "disastrous climatic event"; and second, they examine the ethical, political and psychological impact of climate change on both scientists and non-scientists (2016, 270). Despite scientist characters appearing frequently in climate change plays, these works are generally not grouped together with science plays in scholarship. Hence, while climate change is mentioned by critics as an important topic for both contemporary science plays and science musicals (Shepherd-Barr 2016, 110; David Savran quoted in Wolf et al. 2022, 204), it features little in studies such as *Staging Science* and *Science on Stage*. It is safe to assume, however, that, some of the challenges that face the climate change play also apply to its musical counterpart: the need to "choose types of discourses and aesthetics that will convey scientific knowledge in a manner that raises audience interest" while (potentially) inspiring personal transformation, and the difficulty of representing a phenomenon that is vast in temporal and spatial scale within the specific confines of the theatre (Balestrini 2017, 115). These aspects are especially relevant for works set in the far south.

Antarctica on Stage

While few staged musical works have been set in Antarctica, the south has nonetheless been staged creatively in a wider range of theatrical performances. Previous dramatic productions have used complex sets to depict Antarctica in the theatre. Early examples—such as the London pantomime *Omai* (1785) or the melodrama *The South Polar Expedition* (1841)—showcased representations of ice or icebergs; the latter was staged at Hobart's Theatre Royal to celebrate James Clark Ross's expedition south and included a "splendid ice landscape" and marauding penguins (Leane 2013, 20). More recent productions, in contrast, have equated the emptiness of a stage with the emptiness of the Polar plateau. The German expressionist, Reinhard Goering (*Die Südpolexpedition des Kapitäns Scott*, 1929), used projections to conjure up ice on stage, a technique reprised by Mojisola Adebayo in *Moj of the Antarctic* (2006). Goering's play's subsequent adaptation as *Das Opfer* (1937)—an early operatic representation of Robert Falcon Scott's fateful South Pole expedition (1911/12) with music by 12-tone composer Winfried Zillig—again made use of a minimalist staging, enlisting abstract structures and a chorus of penguins to transport the viewers to the far south. In 1981, the American playwright Ted Tally likewise instructed that "the setting should above all be simple and flexible, close to a bare stage", with bright lights and "towers of sound" used to evoke a sense of the sublime (1981, xi, xvi). Such reliance on sound harks back to Australian Douglas Stewart's acclaimed radio play *The Fire on the Snow* (1944). Stewart approached the challenge of (re) constructing Antarctica by relying on sound, rather than visuals, to transport the audience south. One advantage of this audio approach is the accessibility of the work, as the 'staging' took place over the airwaves, rather than in one specific physical location.

In some cases, playwrights have embraced the physicality of ice to make Antarctica tangible. Howard Brenton's *Scott of the Antarctic: Or, What God Didn't See* (1972)—an excoriating retelling of Robert Falcon Scott's tragic 1911/12 expedition to the South Pole—was set on a commercial ice-skating rink. The physical presence of ice meant the audience had the embodied experience of a chilled Antarctic climate. However, the artificial and mundane nature of the rink, with its clear boundaries, undermined the established narrative of Antarctica as a vast, unforgiving place to be conquered by Polar heroes. Ice also figures on the stage of Lynda Chanwai-Earle's play *Heat* (2008), where the slow melt of a chunk of ice both stands for a fragile Antarctica and parallels the thawing human relationships within the play. *Heat* is unusual in that it focuses on modern scientists, rather than 'Heroic Era' figures from the early 1900s—in this regard it is similar to *Antarctica—A New Musical*. Exploring the manifold aspects of "performing ice", Leane et al. note: "Ice is [...] a particularly revealing substance to perform in or with" (2020, 11). Precisely what it reveals depends on the narrative being told—although ice is materially present on both Brenton and Chanwai-Earle's stages, the same element has very different connotations in these two performances.

'Antarctic' plays are not always set in Antarctica. Manfred Karge's *Die Eroberung des Südpols* (1985) describes "the stage as a stage" (1988, 37) and characters use laundry on a washing line to stand in for their imagined South Pole. The characters (elderly residents of a rest home) in Patricia Cornelius's *Do Not Go Gentle* (2011) actively construct their own versions of Antarctica on stage; Act II takes place within "a labyrinth of crevasses and ice towers" (Cornelius 2011, 10) into which the characters disappear. In such productions, Antarctica is not just a location, but rather an idea around which characters structure either their salvation (Karge 1988) or their demise (Cornelius 2011, Nielsen 2020). Joe DiPietro's off-Broadway musical *Ernest Shackleton Loves Me* (2017) takes a similar approach, focusing on a single mother in contemporary USA, and the premise that the Polar explorer Ernest Shackleton "reaches across space and time to share his heroic journey with her" (Ernest Shackleton Musical). Here, a fridge in the background signals both the domestic setting and the chill of the far south, bringing Antarctica into the everyday. Matthew Knowles's *Shackleton* (2021) was written to be performed outdoors, with a large sheet behind the actors, framing both the stage and the far south as blank spaces to be brought to life by song and narrative. The range of ways in which Antarctica has been staged reflects nuanced attitudes towards the ice-covered continent itself—the same setting can be thought of as vast and empty, complex and sublime, perilous or in peril—and staging choices bring these various approaches to the fore.

Writing and Staging a Popular Antarctic Musical

In October 2016, 175 years after the Hobart Theatre Royal staged *The South Polar Expedition*, the venue hosted the premiere of *Antarctica—A New Musical*. The production was a collaboration between Sundog Productions (comprising book author/producer Bergstrom and songwriter/composer McLaren) and the Tasmanian Theatre Company. Directed by Terence O'Connell, it featured a cast of eight performers and two puppeteers, and was supported by a small ensemble of piano, guitar, bass guitar, violin/viola and cello. The show's run spanned two weeks and had a total audience of over 2200, the largest for an original musical in Hobart. As an 'Antarctic gateway city', Hobart has a high concentration of Antarctic researchers and expeditioners, but the whole city celebrates its relationship with the far south. The premiere preceded the inaugural Australian Antarctic Festival, a now-biennial event in Hobart that celebrates Antarctic science, discovery and connections. The local reception was warm, with one reviewer urging readers to "[f]orget preconceived ideas of musicals, this one takes you on a trip to Antarctica and returns you with hummable tunes" (Ruthven 2016). *Antarctica—A New Musical* allowed Hobartians to experience the continent vicariously and return with icy earworms.

While the 2016 season featured the official premiere of *Antarctica—A New Musical,* the work had been evolving for over a decade (and has continued to evolve since). The idea began in 2002, when Bergstrom, an applied Antarctic ecologist, and McLaren, stopped over for an evening in London between attending an Antarctic science conference in Amsterdam and catching the train to Cambridge and the British Antarctic Survey the next day. After attending a West End musical, McLaren noted that although he enjoyed the evening, he could not remember any of the songs from the production. Confident in his ability to write melodic hooks and knowing that his lyrics generally focus on storytelling, he toyed with the idea of writing a musical. Together, he and Bergstrom identified that their contemporary knowledge of Antarctica created a rare opportunity for theatre.

The basic premise of the plot was to show contemporary life across the year at an Antarctic station—something McLaren had experienced firsthand while employed as an overwintering electrician (Bergstrom's experiences were confined to summers). McLaren had penned a song, *Antarctic*, while wintering at Casey station and had recorded a demo version on Australia's sub-Antarctic Macquarie Island in 1999, after building a makeshift studio in an unused room at the back of the electricians' hut. This song became the central theme. However, it was not until 2006 that significant effort began on the work. After a life-threatening accident, Bergstrom realised she wanted to have a greater impact on the conservation of Antarctica and felt that the musical genre offered the ideal platform for audiences to emotionally connect to the continent. The premise for this approach was that if more people care about Antarctica, there is a greater chance for protection. Over the following decade, the work developed, with workshops in Hobart in 2010, 2012 and 2015.

The initial drafts of the musical's script explored the range of people who live in Antarctica, their motives behind travelling south and the day-to-day activities of a station—from summer through to the darkness of winter and back again. By the 2016 premiere, the plot followed Birdie, a young penguin biologist who heads to Antarctica to expand her horizons. Prior to her trip south, she has a narrow view of the role of scientists in the world, and feels like a fraud when asked to lead. Climate-change-driven extreme events—including a storm that destroys penguin colonies and leads to the expedition's doctor (her romantic interest) and the mountaineer falling into a crevasse while out collecting data for her—push her to challenge these assumptions. Over the course of the musical, Birdie transforms into a hero at the small scale, rescuing people she loves, while realising the world needs rescuing too (as in the song *We Can Rise Above*). This shift of perspective from the local to the global parallels a wider shift in how Antarctica itself has been viewed in recent years—in the age of the Anthropocene, the continent can no longer be viewed as exceptional in its separation from other parts of the world.

As with previous productions set in Antarctica, the problem of how to stage such an extreme place in the interior of a theatre had to be solved. Stage designer Eamon D'Arcy's approach, like that of many previous designers of Antarctic theatre, was minimalist (Fig. 1): a simple white box 3.6 m in height with four doors, two upstage and two on the back wall, parallel to the stage front. Behind the box was a cyclorama, lit to convey mood, with lighting designed by Jason Bovaird. The emptiness of the stage was remarked upon in a review for *Australia Stage*: "The stark, raw beauty of the continent was reflected in the white box set to the point that there was nothing really to look at" (Hern 2016). D'Arcy's vision behind the box was that it was impossible to depict Antarctica on stage because of the scale of the landscape. Ice, which is known for its blinding effects and sublime nature, is also ever-changing, so the white box allowed the audience to imagine their own Antarctica. Ice was therefore communicated through its absence. D'Arcy's approach also paralleled his and Bergstrom's idea that most people's concept of Antarctica is more myth than reality, as so few people have ever set foot on the continent in the span of human history.

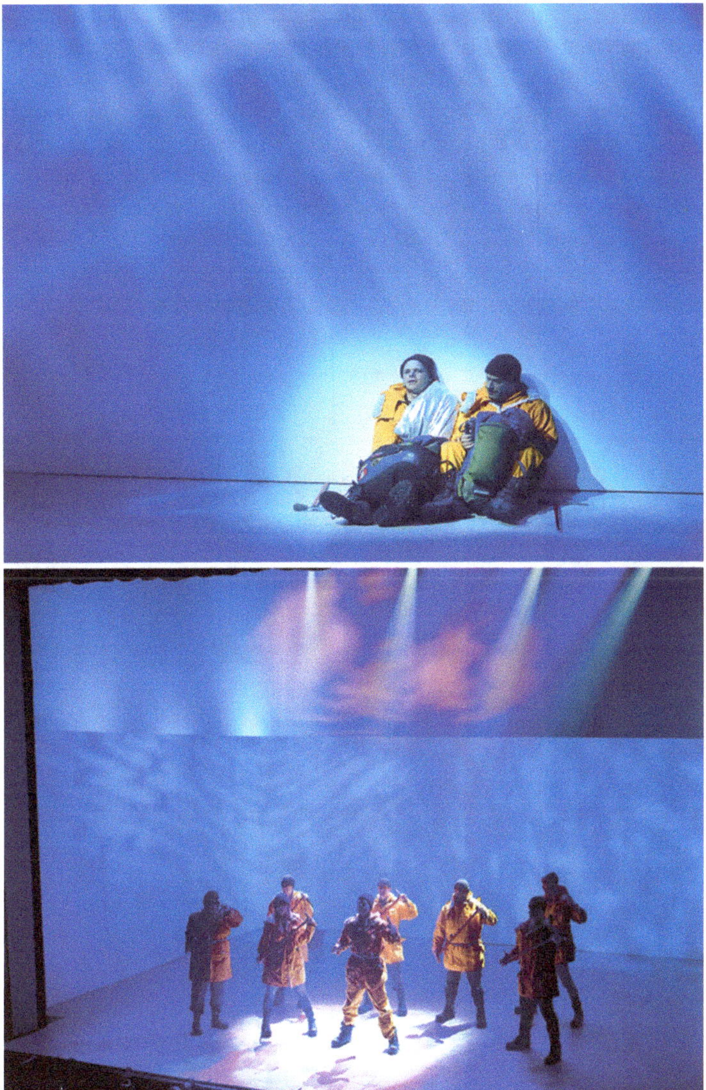

Fig. 1 Richard Parkinson, the minimalist staging of *Antarctica—A New Musical*, 2016 © Sundog Productions

The transition from outside to inside the fictional Hurley Station (named after the Australian expeditioner and photographer Frank Hurley, for whom Bergstrom's mother once worked) was made through the descent of three industrial lights and the wheeling in of one or two tables and some chairs. The tables were also used to signify the deck of an icebreaker in one scene. Scene transitions were made with the aid of actors, but were mainly done by the puppeteers, dressed in white—the reverse of the traditional black Kuroko in Japanese kabuki theatre. In the icy Antarctic environment, white is the colour of invisibility (Fig. 2). The puppeteers symbolise Antarctic spirits who witness the action. They make visible the construction of Antarctica on

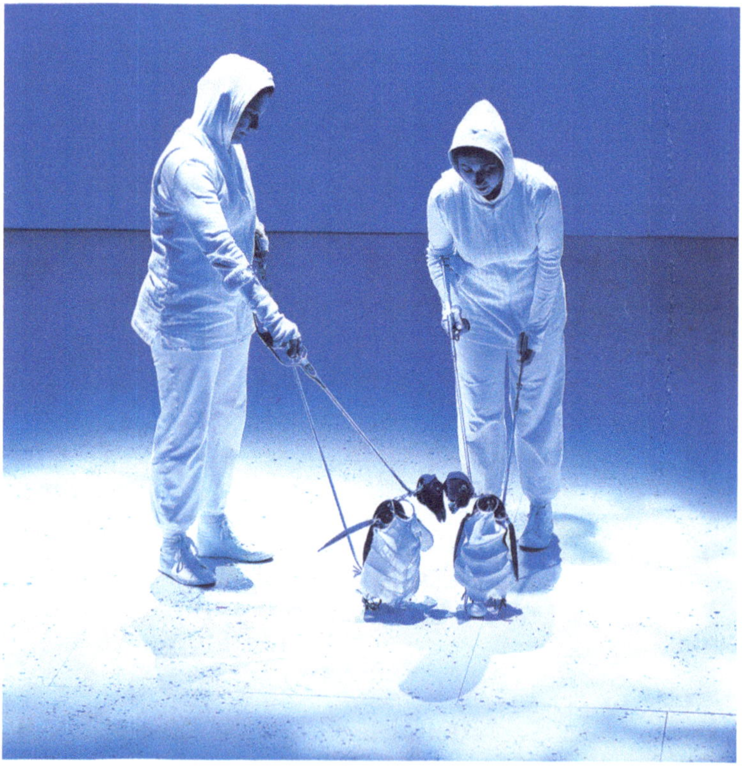

Fig. 2 Richard Parkinson, penguin puppets and puppeteers in *Antarctica—A New Musical*, 2016 © Sundog Productions

stage—echoing earlier productions, such as Karge's—while inviting the audience to actively participate in imagining the place.

The minimalist staging was offset by the use of intricate puppetry—designed and made by McLaren with assistance from Bergstrom—to evoke both vehicles and wildlife. After much experimentation, they settled on an aesthetic that reflected the materials and tools available on an Antarctic station. The puppets were mainly constructed from welded steel or aluminium rods and were primarily covered in translucent fabric. Two scales were used: a miniature icebreaker, helicopter and Hägglunds (an over-snow vehicle) to infer distance, while wildlife—penguins, an elephant seal, a wandering albatross, a giant petrel and a snow petrel—were life-size, interacting with the cast. Supported by Antarctic penguin biologist, Barbara Wienecke, McLaren studied the underlying skeletal features of the animals to guide his designs. Particular attention was paid to the natural movements of the animals, such as the Adélie penguins' flexible spines allowing transparent eccentric wheels to sway the puppet as the puppeteer's left hand operated the body, while the right hand supported the penguins' heads independently. Wings, beaks and head tilt could also be animated to reflect the birds' mannerisms. Finally, a mechanism allowed movement from a squat position to standing. In one scene, a penguin picks up a rock and drops it at the feet of performers and in another, the giant petrel (controlled by both puppeteers) rips at the intestines of a dead penguin, killed during a plot-pivoting storm. Charismatic megafauna, brought to life by visual effects, offer an accessible route to imagining—and subsequently caring about—the Antarctic. As the next section demonstrates, these visual effects were combined with the plot and musical counterparts to complete the transportation of the audience beyond their everyday experiences.

Music and the Antarctic Setting

The music for *Antarctica—A New Musical* was composed in a popular and highly accessible style. The 2016 musical featured 19 original songs (lyrics and music) by McLaren. These songs were arranged for multiple voices and a small ensemble of piano and strings (violin/viola, cello, guitar, bass guitar) by Craig Wood (Musical Director). While the production's musical numbers vary considerably in terms of mood and character, they can nevertheless be divided into four categories based on the thematic content explored: expressions of awe for Antarctica and its environment; music that represents non-human species in Antarctica; songs that relate to science and climate change; and works dealing with human experiences during the expedition. Together, they paint a picture of a wondrous icy place that both inspires sublime emotions and has global environmental relevance.

Three songs in particular, all from the first 30 minutes of the production, establish the story's Antarctic setting by introducing the frozen continent and emphasising the power and inhospitable nature of its environment. The opening number, *The Moment*, features the Station Leader, Elizabeth, singing about her previous experiences of Antarctica. The music begins delicately (the marking on the score specifies "Ethereal") with Elizabeth reflecting primarily on her sensory memories, firstly recalling the stillness and silence: "All is still/My breath the only sound/I can hear above/The snowflakes, drifting down" (McLaren, arr. Wood 2016, unpublished piano/conductor score for *Antarctica—A New Musical*). She then describes her visual memories, including the whiteness and, in the chorus, "Light floating everywhere/Soft and shadow free/Showing more than I/Ever, hoped to see/This is the moment and I/Want to let it wash all over me" (McLaren, arr. Wood 2016). The sparse texture of the mostly consonant accompaniment builds in intensity through the second verse, which concludes with the poignant question, "Why does a world so cold/Bring fire to my soul?" emphasising both the irony and intensity apparent in many human connections to this remote and hostile place (McLaren, arranged by Wood 2016).

Similarly, *Antarctic* and *Nature Rules OK* foreground unique and potentially dangerous aspects of the Antarctic environment, as well as the impact of the region's aesthetic qualities, on the people who visit. The song *Antarctic* describes the ship journey to Antarctica, the crossing of the wild Southern Ocean; sightings of albatross, a first iceberg and whale; as well as human responses to the environment as "awe conquers fear" and one can appreciate "the beauty of silence" (McLaren, arranged by Wood 2016). Arranged for the full company, the song is in the ballad style commonly encountered in musical theatre, while *Nature Rules OK* provides contrast through its upbeat tempo, insistent rhythms, jazz-influenced piano accompaniment and catchy melody line and lyrics—the latter of which convey important safety information about respecting nature whilst working in the field. The Antarctic environment is also represented in various musical interludes (or sections of underscoring) interspersed between the songs and dialogue, cleverly evoking the character of the non-human species brought to life on stage through puppetry, such as seals (accompanied by descending glissandi on low strings) and penguins (which waddle to lively, higher-pitched staccato motives). Through such songs, the musical conveys a sense of the Antarctic experience to the audience—whether the listeners have had direct Antarctic experience or not—in both an engaging and accessible way.

Singing About Science

Science is woven into *Antarctica—A New Musical* in both overt and subtle ways. Homage names honour earlier researchers: the lead character Birdie (Dr Bryony Selkirk) and Bennett's Island are named after pioneering female Australian Antarctic scientists, Drs Patricia Selkirk and Isobel Bennett. The theme of science, particularly relating to climate change, is far more noticeable in two of the musical numbers: *Reality* and *Blue Planet*. As the second song heard in the production, *Reality* plays a key role in introducing the lead character, Birdie. Her enthusiasm for science and her excitement at the opportunity to translate theory into practice in Antarctica are immediately apparent in her lively, fast-tempo delivery of lines such as "I want to test the theory/I want to taste reality/I wanna see, hear, smell and feel/I want to live a life that's real", and especially at the end of the song where she exclaims: "Oh I bloody love science!" (McLaren, arr. Wood 2016).

In contrast, *Blue Planet* in Act II is more serious in subject matter and mood, as female characters Birdie and Skye firmly assert their science-backed position on climate change in an attempt to convince the male characters, Moose and Doug, of the importance and urgency of taking action to address it. Just before the song begins, Moose argues that interests other than science motivate Australia's presence in Antarctica: "Science! There's no consensus in the science! Birdie and her mates toe the line to get grants, and the science down here is a ruse. An excuse to support our territorial claims before we start mining." The juxtaposition of Birdie and Moose dramatises the conflict between an idealistic view of science and a more cynical view that science is infused with politics. Mining is banned under the 1991 Environmental Protocol to the Antarctic Treaty (Madrid Protocol), and the status of all claims to the continent (including Australia's) was 'frozen' by the 1959 Antarctic Treaty, but this passage demonstrates that the view of Antarctica as a place for science and scientists is not shared by all sectors of society. The exchange reflects wider discussions about 'balance' in the media, as Birdie responds angrily to the suggestion that anthropogenic climate change is a ruse:

> With climate change? Sure, there are two stories. One, a reasoned scientific position based on decades of science, by thousands of scientists, complete with peer review. The other, a belief system based on propaganda spewed out by shock jocks, backed by greed and neo-con think tanks and swallowed by the gullible.

This conflict foregrounds questions about who holds knowledge and how it is shared. The responsibility associated with communicating Antarctic science is therefore reflected upon in both the content and the form of *Antarctica—A New Musical*. The pairs of characters soon enter a debate through the song, with the increasing emotional intensity in the delivery of the lyrics matched by the ensemble's accompaniment, which is in a minor key and builds in texture and dissonance as the tension develops on stage. By staging contemporary conflicts over climate change science, the musical explicitly addresses misunderstandings while encouraging reflexivity over the role of the scientist. Birdie's recurring line (sometimes harmonised by Skye), "We can rise above/Blue planet needs our love" (McLaren, arranged by Wood 2016) regularly and positively emphasises the overarching message of care within the song, and this is matched by more consonant, tonal writing in the musical language.

The remaining songs in the production tend to concentrate more on human experiences during the expedition, exploring themes of love and longing, fear and death, the monotony and isolation of overwintering and the anticipation of returning home. Collectively, these songs harness the expressive energy of music to convey some of the many intense and contrasting emotions typically experienced by expeditioners during extended stays on the continent, and simultaneously encourage the audience to feel an emotional connection to the characters, including scientists, and the Antarctic setting.

New Science Questions for *Antarctica*

Since *Antarctica—A New Musical* premiered in Hobart in 2016, new developments have made the environmental issues at the heart of the musical even more pertinent. The fictional rain event that precipitated the deaths of Birdie's penguin subjects in the musical ("the rain has melted the snow and flooded the colony") saw its real-life counterpart in two Antarctic heat waves that saw rainfall around the Antarctic coast and a record high of 18.3 °C on the Antarctic Peninsula on 6 February 2020 (WMO 2021). In Australia, catastrophic bushfires and several one-in-a-hundred-year flood events occurred back-to-back (Bergstrom 2022). Drawing attention to the climate crisis, Swedish activist Greta Thunberg started the *Skolstrejk för klimatet* ("School Strike for Climate") movement in 2018, amplifying youth concerns about the planet's future. Bergstrom's scientific work on heatwaves and ecosystem collapse has continued to inform both scholarship—in 2021 she was awarded the prestigious Eureka Prize for Leadership in Innovation and Science—and the ongoing creative development of *Antarctica—A New Musical*. Reflecting on the initial premiere of the musical, Bergstrom noted: "I don't think there's that much difference between the scientific and creative process" (quoted in Luttrell 2016). Both involve curiosity, experimentation and revision to incorporate new knowledge.

Following solicited audience feedback on the 2016 musical and supported by an Arts Tasmania grant, the work underwent a major rewrite, reflecting international environmental and social developments. However, the COVID-19 pandemic, with its restrictions and lockdowns, necessitated a rethink on the work's format. The most recent iteration of the work is a podcast titled *Antarctica, Beneath the Storm*. Scheduled for release in 2023, the podcast musical brings climate science and the importance of science for Antarctica, to the fore. This version is punctuated with scientific moments ("ARGO float, oohh … ver board!"). It also responds directly to the intervening environmental events, featuring a "one in a hundred-year flood" in Australia alongside a musical reflection on the role of the scientist entitled "It's Not My Job". The new opening—"And the winner of this year's Nobel Peace Prize is … Professor Bryony Selkirk"—positions the scientist as having the political agency and social capital to make real change. The narrator goes on to explain: "When Greta [Thunberg] called on adults to take over so she could go back to school, Birdie was there".

A science play that is responsive to new developments can be considered as a kind of experiment that occurs every time it is performed, as conditions differ each time. From this perspective, we could consider *Antarctica—A New Musical* an iterative experiment that changes in response to reception and feedback, incorporating new data in a manner analogous to any other scientific project. This mode of storytelling is responsive, generative and curious as it opens conversations about Antarctica and the climate with those outside the scientific community.

An Evolving Performance

The shift to an audio format has obvious benefits for a musical set in Antarctica. It removes the need to replicate the Antarctic ice in the confines of a theatre, returning to the medium that Douglas Stewart exploited so successfully in *The Fire on the Snow* (1944). As a podcast, the musical is portable and accessible anywhere in the world, without the expense of staging and touring a production. Its asynchronous delivery widens the scope of the potential audience far beyond the local geographic area. And in the 2022 context, this shift to audio also carries an environmental message—staging a musical requires intensive resources, including electricity/transport, whereas the radio play format is lower-impact. The shift also coincides with a growth in interest in the audio medium more broadly—a 2021 market scoping study showed 62% of the US population were weekly online audio listeners (Edison Research 2021). The audio medium therefore achieves both a lower environmental impact and a greater human impact by turning ears southwards.

In transforming from *Antarctica—A New Musical* to *Antarctica, Beneath the Storm*, the work also changes the focus from relationships between humans to human relationships with the environment. This concept is encapsulated in the refrain "The golden rule is, nature rules OK" and in the two additional songs that centre on environmental themes, further emphasising the science and climate change messaging. "It's Not My Job", inserted at the very end of Act I, features Birdie expressing her frustration about the irrefutable science of climate change continuing to fall on deaf ears ("Proof's been out for years and years/But the suits and ties have coal in their ears"), whilst also asserting that her role as a scientist only goes so far: "It's not my job to stand up and shout it out loud/ … not my job to save you or warn you/ … I build a case, then search for flaws/'cos science seeks truth unswayed by a cause." Similarly, *Revolution*, which is sung by Birdie towards the end of Act II in the podcast version, asserts a strong position on climate change, but is more of a call to arms, with lyrics such as "Greed's not good when it's doing harm/It's time we all took back our planet/… Time has come to drop the acts/To swallow inconvenient facts/The time for revolution's now/No more milking Planet cow", and the repetition of the key message from a few songs earlier: "Blue planet needs our love". This suggests that the job of the scientist is, in fact, broader than collecting data—the stories that data tell and the ways new information is communicated are also of vital importance.

In accepting her Nobel Peace Prize, Birdie reprises the opening of *Antarctica—A New Musical* as she reflects on her relationship with Antarctica: "I still dream of Antarctica. It can be different, different people, a different station, landscape, but it's always the big A./I see echoes of her everywhere."

The podcast *Antarctica, Beneath the Storm* takes these same ripples and spreads them wider, bringing Antarctica to those back home and encouraging curiosity and identification with the icy place. Just as "Climate is much bigger say/than the weather day to day", the importance of Antarctica extends well beyond its continental boundaries. This chapter demonstrates how the popular form of the musical can be enlisted to collapse the distance to the far south. Affective aspects of the music combine with science communication and an urgent message about the need to protect Antarctica from wider human impacts. The development of *Antarctica* from a stage work to a podcast musical is an example of the sort of adaptive, attentive global storytelling that Antarctica's ice and wildlife need.

Acknowledgements The podcast version of *Antarctica, Beneath the Storm* was assisted by Arts Tasmania.

All figures by Richard Parkinson © Sundog Productions.

Bibliography

Balestrini, Nassim. 2017. Cli-Fi Drama and Performance. *Amerikastudien/American Studies* 62 (1): 114–120.

Bergstrom, Dana. 2022. Observing Life on the Edge: Ecosystems as Early Warning Signs. *Griffith Review*. https://www.griffithreview.com/articles/observing-life-on-the-edge/. Accessed 18 Aug 2022.

Cornelius, Patricia. 2011. *Do Not Go Gentle … and The Berry Man*. Strawberry Hills: Currency Press.

DeConto, Robert M., David Pollard, Richard B. Alley, Isabella Velicogna, Edward Gasson, Natalya Gomez, Shaina Sadai, Alan Condron, Daniel M. Gilford, Erica L. Ashe, Robert E. Kopp, Dawei Li, and Andrea Dutton. 2021. The Paris Climate Agreement and Future Sea-Level Rise from Antarctica. *Nature* 593: 83–89. https://doi.org/10.1038/s41586-021-03427-0.

Edison Research. 2021. The Infinite Dial 2021. *Edison Research (Blog)*. https://www.edisonresearch.com/the-infinite-dial-2021-2/. Accessed 16 Aug 2022.

Ernest Shackleton Musical. 2017. Home. http://www.ernestshackletonmusical.com/. Accessed 25 Apr 2022.

Hern, Julia. 2016. Antarctica – A New Musical | Tasmanian Theatre Company & Sundog Productions. *Australian Stage*. https://www.australianstage.com.au/201610318031/reviews/tasmania/antarctica-%E2%80%93-a-new-musical-%7C-tasmanian-theatre-company-sundog-productions.html. Accessed 18 Aug 2022.

Johns-Putra, Adeline. 2016. Climate Change in Literature and Literary Studies: From Cli-Fi, Climate Change Theater and Ecopoetry to Ecocriticism and Climate Change Criticism. *WIREs Climate Change* 7 (2): 266–282. https://doi.org/10.1002/wcc.385.

Karge, Manfred. 1988. *The Conquest of the South Pole and Man to Man*. Trans. Tinch Minter and Anthony Vivis. London: Expression Printers.

Leane, Elizabeth. 2013. Icescape Theatre: Staging the Antarctic. *Performance Research: A Journal of the Performing Arts* 18 (6): 18–28. https://doi.org/10.1080/13528165.2013.908051.

Leane, Elizabeth, Carolyn Philpott, and Matt Delbridge. 2020. Performing Ice: Histories, Theories, Contexts. In *Performing Ice*, ed. Carolyn Philpott, Elizabeth Leane, and Matt Delbridge, 1–21. London: Palgrave Macmillan.

Luttrell, Alex. 2016. Australian Antarctic Division Ecologist Dana Bergstrom Makes the Leap from Science to Musical. *The Mercury*, September 6.

McLaren, Dugald, arr. Craig Wood. 2016. Antarctica – A New Musical. Piano/Conductor Score.

Nielsen, Hanne. 2020. Staging the Construction of Place in Two Antarctic Plays. In *Performing Ice*, ed. Carolyn Philpott, Elizabeth Leane, and Matt Delbridge, 27–54. London: Palgrave Macmillan.

Priestley, Rebecca K, Zoë Heine, and Taciano L. Milfont. 2021. Public Understanding of Climate Change-Related Sea-Level Rise. PLoS One 16 (7): e0254348:1–12. https://doi.org/10.1371/journal.pone.0254348.

Ruthven, Elizabeth. 2016. Rare Gem Is Found in Antarctica Musical. *The Mercury*, November 2.

Shepherd-Barr, Kirsten. 2006. *Science on Stage: From Doctor Faustus to Copenhagen*. Princeton/Oxford: Princeton University Press.

———. 2016. 'Unmediated' Science Plays: Seeing what Sticks. In *Staging Science: Scientific Performance on Street, Stage and Screen*, ed. Martin Willis, 105–123. London: Palgrave Macmillan.

Stevens, Craig, Gabby O'Connor, and Natalie Robinson. 2019. The Connections Between Art and Science in Antarctica: Activating Science*Art. *The Polar Record* 55 (4): 289–296. https://doi.org/10.1017/S0032247419000093.

Sundog Productions Australia. 2021. Art & Science in Theatre: ANTARCTICA, BENEATH THE STORM. 29 Ecological Society of Australia Conference. https://vimeo.com/651049127. Accessed 18 Aug 2022.

Tally, Ted. 1981. *Terra Nova: A Play*. New York: Nelson Doubleday.

Willis, Martin. 2016. Introduction: Imaginative Mobilities. In *Staging Science: Scientific Performance on Street, Stage and Screen*, ed. Martin Willis, 1–9. London: Palgrave Macmillan.

Wolf, Stacy, Masi Asare, Rob Berman, Randall Eng, Eric M. Glover, David Savran, Georgia Stitt, Brandon Webster, and Sarah Whitfield. 2022. What Do We Do with the Musical Theatre Canon? In *Troubling Tradition: Canonicity, Theatre and Performance in the US*, ed. Lindsey Mantoan, Matthew Moore, and Angela Farr Schiller, 201–216. London: Routledge.

World Meteorological Organization. 2021. *WMO Verifies One Temperature Record for Antarctic Continent and Rejects Another*. WMO. https://public.wmo.int/en/media/press-release/wmo-verifies-one-temperature-record-antarctic-continent-and-rejects-another. Accessed 16 Aug 2022.

Icy Love: Performing Affect and Emotion Feeling About Climate Change

Peta Tait

Ice is visually presented as well as verbally described in theatrical performance. Although the staging of ice might seem a particularly effective way of gaining spectator attention, this chapter suggests that an artistic intention to communicate about environmental protection may also need to elicit emotional feeling. It contends that a prosody that moves from seeing (or hearing) to feeling and to caring in performance can enhance viewer participation and awareness of an environmental context (Tait 2022). The discussion assumes that ice in performance has become a short-hand reference to the Anthropocene, climate change and a decrease in Polar ice sheets (Leane et al. 2020)—and therefore implicitly evokes fears about the consequences of global warming and melting ice. Performance with ice may need to counteract resistance to the arousal of anxiety about the climate. This chapter considers examples of twenty-first-century artistic works presenting ice and, in particular, *THAW*, a performance by the

P. Tait (✉)
La Trobe University, Bundoora, VIC, Australia
e-mail: P.Tait@latrobe.edu.au

A. Hemkendreis, A.-S. Jürgens (eds.), *Communicating Ice through Popular Art and Aesthetics*, Palgrave Studies in Media and Environmental Communication,
https://doi.org/10.1007/978-3-031-39787-5_9

Australian physical theatre group, Legs on the Wall, that visually depicted human dependency and struggle with ice. It explores how performance might shift from having a sensory appeal with affects (sensations) of attentiveness to eliciting emotional connection.

The visual presentation of ice can be contrasted with a verbal description in what Elizabeth Leane calls the "icescape" of theatre—in which dramatic characters create a "theatre in the ice" of Antarctica or the Arctic that is imbued with the emotional meaning of heroic endeavours and perceptions of "agoraphobia from the 'vast whiteness' and claustrophobia" in the hut (Leane 2013, 19). Dramatic theatre about ice promises emotional significance, whereas visual art and contemporary performance with ice offer an immediate sensory allure. Arguably, bodily sensations of affect are fleeting while an emotionally felt response can linger in the memory of an event (Tait 2022). Yet, as Marissia Fragkou explains, strategies used in theatrical performance about climate change may not effectively prompt social action unless they encourage imaginative responses to perceptions of precarity and uncertainty (2019, 82–83). *THAW* evoked surprise and curiosity as well as exhilaration in combination with apprehensive concern for the precarity of a performer positioned mid-air on melting ice. It suggested an imaginative convergence of affect and emotional feeling.

Emotionally Evocative

The association of ice and snow with winter and weather in various geographical regions has implications for environmental awareness. Yet local knowledge of changes in ice and snow can be antithetical to climate change conversations—as a study in Norway reveals (Nørgaard 2011). Social attitudes reveal emotional ambivalence and avoidance of the topic of climate in day-to-day communication. As Anne Hemkendreis and Anna-Sophie Jürgens point out (chapter "Communicating Loss: Ice Research, Popular Art and Aesthetics: Introduction," p. 6), Greenpeace activists drew the wider public's attention to change in the Arctic region by collaborating with scientists and artistic practitioners (2023, 2). This type of collaboration suggests that scientific information needs to be supported by emotional engagement, which artistic works can encourage.

The qualities of ice in the landscape vary greatly according to its geographical location and ensuing cultural depictions (Leane et al. 2020). Attitudes can be contradictory; icy weather is dreaded even as snow-covered vistas are celebrated for their aesthetic appeal. Jonathan Pitches explores some of the attractions of mountains and snow-covered peaks as he observes that humans are increasingly "treating mountains as a performing landscape" through the insistent filming of activities with views (2020, 9). Pitches contends that art, film and live performance communicate the "visceral, visual and vertiginous experiences" that shape human responses (2020, 11). He analyses how a landscape of mountains is increasingly overlaid with specific attributes including the contribution of humans to the melting of ice. An image of mountain ice and snow is viewed through layered cultural perspectives. In addition, Pitches explores how performers within diverse cultures embody concepts of mountains; he finds metaphors—for the extreme effort of walking and climbing with apparatus—within performer training and in performance itself. Physical effort overlaps with emotional effect in performance. Recent technological and economic developments, however, allow pedestrians to approach the heights of snow-covered vistas with comparative ease in places—without the physical effort of a mountaineering climb—on skyways and skywalks that mark landscapes with features of human control (Pitches 2020, 220). Mountain vistas are being objectified and consumed in performative processes (Pitches 2020, 229). The trepidation, uncertainty and fear once denoting mountainous regions and climbing are being displaced in social media; they are being framed with human hubris.

Contemporary performance can reinvigorate the longstanding fearful reverence for ice in the cultural imaginary. VestAndPage (Pagnes and Stenke 2013) performed on and in ice, including in Antarctica, and created artistic images of human bodies that elicit bodily sensations of affect in combination with emotional feeling. An image of a naked male and a naked female in an embrace surrounded by green ice juxtaposes extreme metaphors of heat and cold as it extends the idea of human love to the ice, and the emotional riskiness of lovemaking converges with that of skin exposed to deadly cold (Pagnes and Stenke 2013). This image, associated with "Ice, Pure Oxygen and Penguin Shit", emerges from a trilogy of films called *sin∞fin The Movie* (Pagnes and Stenke 2013; Quinn 2020). Another equally evocative image surprises as it disturbs: a fleshed hand holds a human-shaped hand made of ice; the hand-holding evokes impressions of tenderness, love and climatic extremes.

VestAndPage are well known for these site-specific performative works undertaken live in Antarctica; the author viewed their filmed work at the first International Performance Week, Palazzo Bembo, an event which the artists also curated (Venice, 8–15 December 2012). The aesthetic effect of a distant human figure lying or standing in an icy landscape is captivating and imbues the landscape with a sense of mystery. VestAndPage describe the "beauty and danger"—reverence and fear—of performing live actions on the ice in Antarctica for the camera, where ice in Antarctica "is more real than reality itself"; time flows differently because humans can only withstand limited exposure (2013, 72). Their artworks convey fragility within ideas of human love, and love for ice. VestAndPage contend that the Antarctic continent should not be a place for humans—reflecting their hope for its preservation as a pristine nonhuman domain more than a pronouncement of an inhospitable world. The imagery also implies that its icy fragility is metaphoric of all environments on which humans depend.

It seems, however, that it is nonhuman animals (henceforth animals)—such as the (emperor) penguin (Leane et al. 2020) or the Polar bear—who galvanise widespread public expressions of care and concern for the preservation of icy regions. Polar bears, in particular, embody climate change in the popular imagination. Una Chaudhuri (2012) explains that one particular photograph in 2006 suggested that a Polar bear had become stranded on a section of floating ice and thus became an international emblem of decreasing northern hemisphere ice sheets. A forlorn bear sparked heartfelt concern for the Polar regions and embodied climate change for the general public—even though the actual location was subsequently revised. Chaudhuri explains that high-profile American politician, Al Gore, drew attention to the photograph saying that the bear had "nowhere else to go", which galvanised mass media attention to ideas of place (2012, 46). Gore highlights what Chaudhuri calls a "geopathology": anguish over dislocation from place and the geographical displacement of other species (Chaudhuri 2012, 46). Ice is place-specific and emotional concern for a vulnerable bear highlights a geopathological destruction of other animals' places.

Chaudhuri contrasts the artistic presentation of taxidermied Polar bears in iceless domestic settings with a live performance in which a human performer replicated the conditions of a popular Polar bear in a zoo. The former garners affects of strangeness, even the uncanny, while the latter evoked emotional distress and compassion. *Polar Bear God* highlighted the plight of popular Polar bear, Gus (in New York's Central Park Zoo), when artist Deke Weaver performed in a small space equivalent to Gus's walled case. The performance conveyed impressions of cruelty and the inflicted misery of a captive life. The human performer took on the dimensions of the bear's confinement and physical "repetitive rhythms": "moving blindly" between rock wall and glass wall in "stereotypies"—when animal behaviour reflects trauma and "zoo psychosis" (Chaudhuri 2012, 53). The impact of Weaver's physical performance and the photograph of a stranded Polar bear prefigure a continuum of artistically evoked emotional responses. A human or nonhuman animal presence seems likely to attract emotional engagement.

Ethics of Wonder

The novelty of ice in art and performance might seem a justifiable way of encouraging public action on climate change, but is it effective without an embodied presence to stir an emotional response? In charting how "icy" has become part of eco-art while querying the ethics and efficacy of exhibiting ice by itself, Simone Hancox focuses her analysis on Olafur Eliasson's *Your Waste of Time* from 2006, in which he collects six tons of ice from a receding glacier lagoon in Iceland and transports it to a freezer room in Neugerriemschneider Art Gallery (Berlin). Hancox interprets Eliasson's installation as offering a performative encounter with ice; she is asking if tons of ice in a gallery can destabilise the binaries of nature and culture and human and nonhuman and provoke viewers to mitigate their "environmental impact in the future" (Hancox 2013, 55). Hancox queries the limitations of such art in encouraging social action, finding that while the creation of *Your Waste of Time* itself was carbon-using and wasteful, it "might" have had a myriad of thought-provoking effects, depending on the level of knowledge brought to this "form of non-didactic eco-awareness" (2013, 55–6). Above all, however, she finds that the unique shapes were "beautiful", shimmering white with tinted hues in the gallery space. In other words, the sensory and physiological affect of viewing ice in art and performance can be a compelling aesthetic experience. I have had similar responses to those of Hancox: ice in performance is beautiful, and a cognitive ecological meaning is not necessarily inherent—even when humans handle it.

In describing affective fascination, Jane Bennett explains that phenomenological feelings of enchantment can lead to "enjoy-ment over pity", which does enhance a capacity to be ethical (2001, 12, 13). She draws on Nietzsche to contend that "one of the tasks proper to ethics is to 'en-joy' the world" (Bennett 2001, 13). Bennett is suggesting that an emotional experience—such as enjoyment—encourages a fuller appreciation of an ethical issue. Bennett's affective fascination can lead to enchantment, and it can be applied to the aesthetics of ice and its beauty within responses ranging from distraction to wonder at its dazzling qualities. The ethics, in this instance, concern the "proper" taking of action to forestall melting due to climate change.

A performance offering a clear environmental message was Latai Taumoepeau's durational *Repatriate II* (2015, Sydney)—the author viewed this outdoors at Carriageworks—in which her white-suited figure shovelled piles of ice. In *Repatriate I*, Taumoepeau performed in water to draw attention to how rising sea levels are destroying Pacific islands and their cultures (Tait 2018, 2022). In *Repatriate II*, her actions in an iceless, snowless Sydney suggested uncertainty and precarity by confounding assumptions about the stability of the climate of a particular place. The question as to whether this type of performance will encourage ethical social action to preserve nonhuman environments remains. Yet, it seems that bodily engagement and emotionally felt responsiveness to place, facilitate unifying tendencies that can override human separation from the nonhuman world and encourage greater environmental and climate awareness (Tait 2022). The specificity of place encourages its preservation. Even so, effective action about place requires that the affect and emotionally felt ambivalence behind inertia and disengagement need to be challenged (Lertzmann 2015).

Hancox draws on Peggy Phelan's account of the "ontology of performance" as disappearance in relation to art about ice, and to suggest that melting ice has performative agency—irrespective of performer presence. This performative concept is illustrated with Néle Azevedo's *Minimum Monument* (*Melting Men*), which the artist began exhibiting in 2005. It consisted of 1000 small, icy humanoid figures that gradually melted on steps in public space; this international exhibition highlighted human dependency on frozen lands (Hancox 2013, 54). The fragility and precarity of this installation also invited an imaginative association of humanity disappearing with the ice. Artistic images of these endearing diminutive figures evoke pathos and loss as they raise questions about survival for human observers.

Legs on Ice

THAW, by Legs on the Wall (henceforth Legs), was both a conceptual performance about ice and a sensual experience of ice, within a human-to-nonhuman dynamic that stirred a range of emotional responses, including enjoyment. An aerialist, suspended in the air, was standing on, flying across and pushing out from, a large triangle of green-blue ice elevated over sea water. *THAW* involved nonverbal, solo mid-air action performed over eight hours by a succession of female performers (Vicki van Hout, Jenni Large, Victoria Hunt and Isabelle Estrella) on a two-ton block of ice positioned in Sydney Harbour. Directed by Joshua Thomson, it had an accompanying evocative aural soundtrack composed by Matthew Burtner (from Alaska), and was staged above and to the side of a pedestrian walkway (Legs on the Wall online 2021); it could be viewed live by standing below or through a filmed version (viewed by the author 15 January 2022). The ice block had one flat upright surface and a pointed bottom so that its shape—not the carbon-less ice—was reminiscent of a diamond and compressed carbon; the dripping fragility of ice provided a motif of continuous release.

THAW's lone human figure reached out to the ice in temporal momentum. Standing on the ice mid-air, a female performer moved rhythmically, her legs and arms spreading out as she flew up into the air. Swinging her body outward, she attempted to hold on to the ice, as if in an embrace. Standing, she seemed to struggle to stay on the ice as it tilted beneath her; it dripped below her feet. She lay down on it, running her hands over its icy form, the heat of the human body contributing to the melting, slippery surface. She flew outward into the air, circling around the flat top of the ice, pushing her feet off the sides, swinging and returning back in gentle arcs of aerial motion.

Anne Hemkendreis (2022, 2) describes how *THAW* conveys an idea of an "individual struggle" with ice, and, therefore, with climate change, finding the performance "haunting" in the way it combines the "sublime luminosity" of ice and kinaesthetic impact of aerial action. She notes a Greta-Thunberg-style costuming in one persona. Hemkendreis finds that the performance was emotionally evocative and contextualises it in relation to theatrical and music performance about, and with, ice. Her extended analysis brings to the fore how there is science and technology behind understanding ice and undertaking aerial action. She reiterates that it is the displacement of human dominance—and emphasis on the dependency of the human on the nonhuman world—that makes this type of art crucial.

THAW was a technologically mediated visual spectacle with the performer in a harness attached to aerial rigging and the ice suspended by four metal cables attached to a large crane, positioned parallel with the outdoor walkway on the west side of the Sydney Opera House building. It was performed on 14–16 January 2022 as part of the Sydney Festival. Towards the end of the show, the performers unfurled a banner that read, "No Time to Waste". The performance happened outside the most iconic performance venue in Australia, giving *THAW* and its message an ironic association with its culturally important "high-art" stages.

As an affective experience, *THAW* was fascinating. The performer was in constant motion. It implied a sense of urgency in connection with the ice, conveying an "*active* quality that ice shares with humans" since it is "dynamic: mutable, unstable, mobile and constantly in transition" from hardness to fluidity (Leane et al. 2020, 2, italics in original). The movement was a mixture of awe-inspiring aerial dance and choreographically risky feats, shaped by the limits of suspended ice. The musical soundscape of *THAW* was lyrical, rising and falling like the sound of the sea below, shifting in and out of a listener's conscious awareness. On camera, the ice visibly dripped into the sea so that the performance became a literal version of Polar ice dissolving into rising sea water—an implicit suggestion of Antarctica. The sea below is part of a working harbour—connected to ocean globally and fossil fuel-burning shipping appeared in the background. Seagulls flying and calling around the performer seemed like an airborne audience in the shared space of the atmosphere.

I can attest to my own enjoyment of the aerial performance. My initial response was one of curiosity about how the performer was positioned on the ice. Her mid-air action invited attentiveness and her speed and flailing movements suggested that she was not in control. Was she stranded? The ice drew my focus and sensory effects oscillated between a haptic sense of coldness and the exuberance of a body in weightless aerial action. The action supported by the soundtrack heightened the exhilaration, as the performer swung out and up repeatedly, circling away from and back to the ice, mid-air. A viewer (like myself) might have become apprehensive about its riskiness, watching with concern for the performer's safety. I was aware of my teeth closing—in response to the cold—when a performer lay down across the ice block, and phenomenological perceptions escalated from seeing to feeling to caring about the performer in an emotional prosody.

This complex performance comes from a 40-year-old company renowned for politically innovative, nonverbal physical theatre. The work of this Sydney-based acrobatic and aerial company has inspired me since 1990, when it began to consolidate its brand of indoor and outdoor performance. Legs revealed that physical bodies in action could be queer and funny, and sad and tragic. A claim that nonverbal physical theatre can evoke pathos might surprise some, but in *All of Me*—written by Mary Morris (a dramatist) and directed by Nigel Jamieson in 1994—Legs explored the loss-filled spaces of a relationship break-up and its impact on children (The Australian Live Performance Database online). A performer seemed child-like in high acrobatic dives, as bodies in constant acrobatic motion conveyed emotional dislocation. While the family break-up scenario is commonplace in theatre, the physical theatre form offered special insight into the way a family exists in continuous body to body contact, which means an emotional break-up also involves wrenching physical separations. The physicality of emotional feeling became apparent.

Outdoor performances by Legs have long contained environmental significance because the company expanded working with vertical height in extraordinary action down the outside of multistorey buildings. The performances defied the way the corporate cityscape towers over humans, making them diminutive. Company members deployed acrobatic physical skills to abseil down the side of high-rise buildings—that is, down mountain-like urban structures—evoking admiration and fears for their safety. For example, members of Legs came down Sydney's Australia Square building in their show *Exbendable* for the 1996 Sydney Festival. In contrast, on the ground, performers became engaged in acrobatic dynamics with each other, as if contesting the confinement of ground-based life. During the 2000 Olympic Games Festival in Sydney, Legs revived *Homeland* (about migration, directed by Debra Batton) by performing down the side of the old ten-storey AMP insurance building at Circular Quay, which had imagery projected onto it (The Australian Live Performance Database online). During the 2006 Commonwealth Games, Legs performed down the side of the National Gallery of Victoria, moving within video projection on its wall, which placed human figures inside geometric patterns and three-dimensional imagery, as if inside a technological (game) world. These outdoor performances involved performers descending and swinging out on wires in slow motion and graceful poses, so that they became bodily silhouetted against the sky in tantalising impressions of boundless action and unrestricted physical freedom.

The company's suspension of bodies between sky and ground can be reinterpreted for its environmental significance. By implication, earlier outdoor works confronted the impervious effect of the high-rise buildings that house corporations and their political worlds and uphold economic growth on a finite planet. (It should be noted that cement adds a significant proportion of carbon to the atmosphere, International Energy Agency 2009.) Legs made the precarity of humans visible within city spaces, the physical risks pointing to some larger metaphoric riskiness for humanity. *THAW* continued the Legs practice of making gravity-defying mid-air performance politically meaningful by highlighting human-made environments.

The performer on the ice provided a potent reminder of humanity's dependency on the nonhuman world and the struggle to halt the threat of rising sea levels from melting ice—and the threat to the famous Sydney Harbour. Did *THAW* encourage more concerted environmental action among spectators? Possibly. The performance vividly reminded viewers that time has been wasted and is running out, with action compounding feelings for ice in a celebrated (cultural) place that stirs emotional responses. The aerial performance engaged the senses in vibrant and visceral ways as the performer's interaction with ice gradually shifted to tenderness. These impressions linger in the imagination.

Prosodies of Love

Near the end of *THAW*, a performer waved to the audience and seemed to make a heart shape with her hands, a sign of love. The gesture suggested a performer communicating with the audience as well as a feeling for ice. It asked for feeling in return. As this performance progressed, it unfolded a prosody of seeing then feeling then loving; performance enjoyment can evoke strong feelings of affection. This type of performance in this important outdoor environment echoes a sequence that arises in relation to place.

An escalation in embodied engagement may motivate environmental action. Nico Frijda connects motivation and motivating action to the concept of prosody—used to describe changes in music and speech that arise from "differences in speed, changes in speed, in the forms of change such as steepness or gradual growth"—as he applies prosody to emotion and emotional feeling (2004, 160; Tait 2022). Prosody can be illustrated with embodied variation that ranges from changes in breathing patterns to changes in physical behaviour, and it can be applied in performance in graduated shifts that, for example, encompass "stirring, arousing, inspiring" (Tait 2022). This offers a way of acknowledging and describing viewer responses that are active and fluid, unfolding through an embodied viewing of a performer in constant motion.

The ways in which performance impacts on the bodily affect of a viewer become more apparent with physical theatre. While a slight bodily tension in the viewer might pass unnoticed, the activation of curiosity and excitement are noticeable, and aerial action often causes a responsive jolt that highlights the sensory process of watching performance. The concept of prosody conveys how bodily responses might have momentum as intended by performance and its music. Prosody denotes how performance can be participatory and motivating as it escalates from telling to prompting to motivating, and a stronger graduation might be that of persuading to inciting to compelling (Tait 2022). The prosody of performance unfolds towards effective engagement.

Performance that evokes bodily feeling and solicits care and concern—for a performer, a nonhuman entity and a place—may encourage action on their behalf. It activates emotional engagement (even love) in a human-to-nonhuman relationship and our crucial dependency on ice. In addition, physical theatre with ice such as *THAW* highlights how risk and danger, precarity and uncertainty can be present alongside feelings of exhilaration and admiration. It emulates the mixed-up feeling of lived experience that can overcome inhibition and inspire social action to forestall melting ice. *THAW* encourages recognition of surprising, struggling and striving in the prosody of gravity-defying performance, as it also confirms a human capacity to defy material limitations and reject perceptions of containment and restriction.

Bibliography

Bennett, Jane. 2001. *The Enchantment of Modern Life*. Princeton: Princeton University Press.

Chaudhuri, Una. 2012. The Silence of the Polar Bears: Performing (Climate) Change in the Theater of Species. In *Readings in Performance and Ecology*, ed. Wendy Aarons and Theresa May, 45–57. New York: Palgrave Macmillan.

Fragkou, Marissia. 2019. *Ecologies of Precarity in Twenty-first Century Theatre*. London: Methuen Drama.

Frijda, Nico H. 2004. Emotions and Action. In *Feelings and Emotions*, ed. Antony Manstead, Nico H. Frijda, and Agneta Fischer, 158–173. Cambridge: Cambridge University Press.

Hancox, Simone. 2013. The Performativity of Ice and Global Ecologies in Olafur Eliasson's Your Waste of Time. *Performance Research* 18 (6): 54–63. https://doi.org/10.1080/13528165.2013.908057.

Hemkendreis, Anne. 2022. Ice as an Agent in the Aerial Performance "Thaw". *Between Science and Art*. https://doi.org/10.55597/e7873.
Hemkendreis, Anne, and Anna-Sophie Jürgens. 2023. Communicating the Invisible: Ice, Research and Culture – Introduction. In *Communicating Ice Through Popular Art and Aesthetics*, ed. Anne Hemkendreis and Anna-Sophie Jürgens. Cham: Palgrave Macmillan.
International Energy Agency. 2009. *Cement Technology Roadmap: Carbon Emissions Reductions up to 2050*. Paris: OECD Publishing.
Leane, Elizabeth. 2013. Icescape Theatre: Staging the Antarctic. *Performance Research* 18 (6): 18–28. https://doi.org/10.1080/13528165.2013.908051.
Leane, Elizabeth, Carolyn Philpott, and Matt Delbridge. 2020. Performing Ice: Histories, Theories, Contexts. In *Performing Ice*, ed. Carolyn Philpott, Elizabeth Leane, and Matt Delbridge, 1–25. Cham: Palgrave Macmillan.
Legs on the Wall. 2021. Home. Legsonthewall.com.au. Accessed 2 Nov 2022.
———. 2022. THAW (Program). https://www.legsonthewall.com.au/thaw. Accessed 30 Nov 2022.
Lertzmann, Renee. 2015. *Environmental Melancholia*. London: Routledge.
Nørgaard, Kari Marie. 2011. *Living in Denial: Climate Change, Emotions and Everyday Life*. Cambridge, MA: The MIT Press.
Pagnes, Andrea and Verena Stenke. (VestAndPage). 2013. Antarctic Dream – Ice as Architecture of the Human Spirit. *Performance Research* 18 (6): 71–80. https://doi.org/10.1080/13528165.2013.908059.
Pitches, Jonathan. 2020. *Performing Mountains*. London: Springer.
Quinn, Douglas. 2020. Figures in the Landscape. In *Performing Ice*, ed. Carolyn Philpott, Elizabeth Leane, and Matt Delbridge, 55–86. Cham: Palgrave Macmillan.
Tait, Peta. 2018. Site-specific Ecological Loss. *Performance Research* 23 (4/5): 192–194. https://doi.org/10.1080/13528165.2018.1522135.
———. 2022. *Forms of Emotion: Human to Nonhuman in Drama, Theatre and Performance*. London: Routledge.
The Australian Live Performance Database. AusStage. https://www.ausstage.edu.au, Accessed 25 Oct 2022.

Pop Cultural Meanings of Ice in Visual Fiction and Film

Frozen Balloons: Aeronautic Heroism and Scientific Knowledge Production

Anne Hemkendreis

Aeronautic Research on Ice and Ecological Awareness

The film *Aeronauts* by Tom Harper (director) and Jack Thorne (writer) is a biopic of the British meteorologist James Glashier (1809–1903). He is considered a pioneer of meteorology and aerology. His scientific merit lies in the investigation of the stratosphere—which was largely unknown at the time—and in bringing the planet's biosphere to people's attention. The film is about Glashier's spectacular flight with the giant gas balloon *Mammoth* on 5 September 1862 from Wolverhampton (UK). On that day, Glashier and the aeronaut Henry T. Coxwell (1819–1900)—replaced in the film by the fictional character Amelia Wren—ascended over 11,000 m (36,000 ft), which was a world record at that time. The balloon flight went down in history less for its scientific results than for its

A. Hemkendreis (✉)
University of Freiburg, Freiburg im Breisgau, Germany
e-mail: Anne.hemkendreis@sfb948.uni-freiburg.de

A. Hemkendreis, A.-S. Jürgens (eds.), *Communicating Ice through Popular Art and Aesthetics*, Palgrave Studies in Media and Environmental Communication,
https://doi.org/10.1007/978-3-031-39787-5_10

spectacular character, which almost caused the death of the scientist and his aeronaut. The icy air frosted the balloon's valve through which gas could be regulated, leading to an uncontrolled ascent that could only be averted by Coxwell's daring action. While Glashier lost consciousness, the aeronaut climbed into the upper ring of the balloon and pulled the rope of the valve with his teeth, as his hands were frozen. In contrast to history, the film shows an even more dramatic rescue, in which the female protagonist saves her and Glashier's lives by climbing up the outside wall of the frozen balloon and forcing the valve open. In doing so, the film turns history into an emotional and dramatic spectacle in icy atmospheres. It further discusses the history of meteorology and its protagonists, the practices of knowledge-making with ice and their mediations.

Snow and ice were central to the life of James Glashier. The real-life, historic meteorologist researched the unique structure of icy phenomena by placing glass plates on a windowsill and catching snowflakes to study under a microscope (Chichester 2021). This resulted in beautiful illustrations of crystallised ice in his book *Snowflakes*, published in 1863, which effectively communicated atmospheric knowledge via an aesthetic experience. As recently noted, Glashier's "rough structural sketches"[1] were completed by his wife, Cecilia Glashier (underrepresented in the history of snow science), by using drawing instruments and taking into account James Talbot's photogenic process (Chichester 2021, 18). Cecilia Glashier's transformation of filigree ink drawings on light-sensitive paper into ornamental line formations suggests a "self-imprint of a natural truth",[2] i.e. an immediate evidential experience of nature through visual representations (Chichester 2021, 18). James Glashier himself considered a sensual address—as clearly emphasised in an article from 1857—to be essential in reaching a broad audience. His book also fostered collaborations between the natural sciences and applied arts (Chichester 2021, 19). In the film, the book on snowflakes marks a turning point within the plot. When Amelia Wren doubts her participation in the expedition due to her personal trauma—her husband jumped to his death from a gas balloon to save her some years ago—it is the beauty of the printed snowflakes that reminds her of the alluring realm above the clouds. While her decision to join the expedition is made, snowflakes suddenly fall from the sky, adding a fateful dimension to Wren's new role as an aeronaut.

[1] "Grobe Strukturskizzen", all translations from German by the author.

[2] "Selbstabdruck einer Naturwahrheit."

At the beginning of the film, James Glashier explains the scientific reasons for his ascent into the iciest layers of the atmosphere in a speech to the Royal Society, which consists solely of elderly men (as was the custom in nineteenth-century science societies). He expresses his wish to explore the earth's magnetic field and the solar spectrum, to gather knowledge about dew points and advance the understanding of the oxygenation of the atmosphere. The phrase "to find order in the chaos" sums up the guiding principle of his research. However, science is not staged as a pure end in itself. After being rejected by the Royal Society, Glashier turns to the widow Wren to explain the ethical reasons underlying his daring plan: "To understand the weather, Miss Wren, is to understand how to make ships and sailors safer, farmers more productive, so we can prepare ourselves and our world for floods, for droughts, famines. We could save thousands of lives." Apart from the possible productivity-enhancing effect for the agricultural economy, the goal of preventing natural disasters sounds less like a need of the industrial age than one of the twenty-first century. Indeed, the speech underlines, from the outset, the film's relevance to the present in terms of the rising threat of climate change. Early meteorology is marked as the cradle of today's climate research, its subject areas and goals. It is further stressed by Glashier's second speech (to the Royal Academy), after his spectacular flight, when he claims a successful "advancement of knowledge" to the general "good for us all" as his ultimate merit. Amelia Wren, in comparison, seems to directly address the viewer with her final words: "You don't change the world simply by looking at it. You change it through the way you choose to live in it." The heroic undertone of these phrases and their empowering gesture go hand in hand with the ecological appeal of the film.

To stress the ecological framing of the biopic, the movie uses cinematographic techniques to highlight a destabilised perspective of humankind, both when the basket of the balloon is shaken by rough weather conditions and when featuring rural landscape scenes (Fig. 1). For example, the final scene iconographically follows romantic traditions by showing a landscape with small figures which flattens out at the edges (as often seen in the work of the painter Caspar David Friedrich). According to Eva Horn and Hannes Bergthaller, the roundness of romantic landscape scenes and their atmospheric skies challenge anthropocentric (human-centred) perspectives (2019). As the authors argue, since Romanticism, the curvature of a landscape scene conveys a feeling for the earth as a planet. Landscape is visualised not as a solid stage for human activity (as, e.g., in the tradition

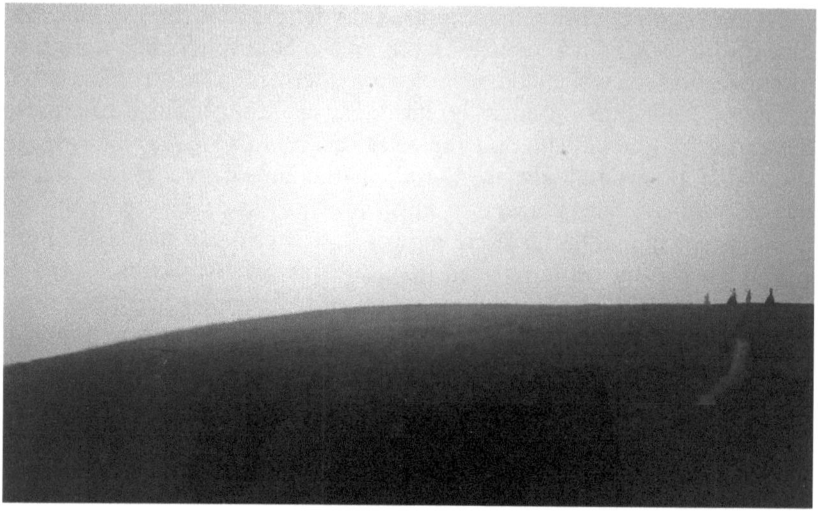

Fig. 1 Harper, Tom, *Aeronauts*, Mandeville Films/Film Nation Entertainment, film still, 2019

of painter Nicolas Poussin), but, instead, evokes the feeling that human's perception is limited when facing the powers of nature. In addition, rejecting a clear vanishing point—as a safe standpoint for the viewer—in favour of an immersive impression stirs feelings of uneasiness and insecurity. Thus, the film uses visual themes that suggest that "in the Anthropocene [...] nature can no longer be represented as a stable given, but remains non-totalisable and non-objectifiable" (Horn and Bergthaller 2019, 118).[3]

The depiction of nature in the film is changeable, fragile and full of beauty; there is a mesmerising swarm of yellow butterflies, an engaging light apparition which doubles the balloon within an aureole and constantly changing clouds. The earth and its atmosphere are presented as a living organism, both familiar and foreign to the protagonists (Elias and Moraru 2015). High up in the air, Glashier and Wren experience "new possibilities of ecological awareness [...] that are increasingly detached from their anchorings in particular geographies" (Elias and Moraru, xii). Their insights into nature are not limited to their geographical departure

[3] "im Anthropozän (...) Natur nicht mehr als stabile Gegebenheit darzustellen (ist), sondern nicht-totalisierbar und nicht-objektivierbar verbleibt."

point in Wolverhampton, but have a more meteorologically comprehensive and global value with regard to the atmosphere. The general beauty of nature in the icy layers of the atmosphere is both the drive for the expedition and, as will be discussed, the cause of interruptions and hindrances to Glashier's meteorological recordings. In a central scene of the film, the scientist reflects that while he derives comfort in his scientific objective of finding the order in chaos—"we may be able to explain the science behind an aureole or the falling snow"—but it is "not possible to account for [its] beauty".

Sublime images of the balloon in a sea of clouds overwhelm the actors in the film and sensually affect the audience. The revival of the sublime has been widely discussed in terms of how it evokes environmental consciousness and a sense of connectedness to the world (see e.g. Brady 2013). The sensuous attraction of sublime images transports the message of one's own physical entanglement with the world into a physical experience, and can liberate the recipient from self-centred attitudes (Bloom 2022; Boetzkes 2010). At the same time, however, the cinematic sublime is suspected of having a passivating, entertaining effect that stands in the way of fact-based gains in knowledge and (paradoxically) an enhancement of subjectivity (Schneider 2021). Furthermore, the sublime has been under a general ideological suspicion due to its use in political propaganda, which is why the connection between aesthetic experience and ethical behaviour remains unclear (Gründler 2019). Thus the sublime can be used for knowledge transfer but it can also be mis-used for spreading fake information. The film *Aeronauts* plays with a decided tension between overwhelming stagings and fact-based knowledge communication in the history of balloon flights, showing that both sides have always been intrinsically connected.

Icy Spectacles and Aerial Heroism

The balloon results from the epistemic and technical achievements of the eighteenth century (Wissmann 1969). However, its true history starts (more or less) with the invention of the hot-air balloon by the Montgolfier brothers, who were the first to raise an unmanned balloon to an altitude of 2000 m on 5 June 1783 (Wissmann 1969, 81). Inspired by Joseph Priestley's publication *Experiments and Observations with Different Kinds of Air*, the scientists experimented with the behaviour of hot air and initially believed they had discovered a new gas. On 19 September 1783, they launched their first fabric balloon into the sky above Versailles with a rooster, a sheep and a duck as its crew. The first balloon flight with humans took place a little later, on 21 November 1783. Since gas expands at colder altitudes, it was not until the invention of the gas valve—making it possible to lower the balloon safely—that crewed scientific expeditions could be undertaken.

The study of weather was for a long time not regarded as a scientific discipline (Gamper 2011). In the film, this is stressed when a Royal Society member describes Glashier's scientific endeavour as an attempt at clairvoyance. Robert Hooke's publication *Method of Making a History of the Weather* from 1667 and Robert Boyle's *General History of Air* from 1692 are early examples of a young scientific discipline trying to stand its ground. However, until the early nineteenth century, weather was not recognised as a continuous and global phenomenon; this changed with publications from Jean André Deluc and Alexander von Humboldt (Wissmann 1969, 404). Ultimately, studies of the weather and atmosphere needed aeronautic travel to open up unimagined possibilities; the balloon became a "technical vehicle"[4] which could take instruments far higher than mountain peaks (Wissmann 1969, 405). It enabled researchers to personally experience the psychophysical consequences of being in icy atmospheric layers, making researcher bodies part of the challenging experiments. Balloon flights were therefore an experimental undertaking in two senses: On the one hand, they were a "place of scientific measurements and experiments that had a function within the framework of a systematic overall explanation of the atmosphere and the processes in it", and on the other hand, they were an "advance into an area unknown to man, which challenged him physically and mentally and drove him to his existential limits" (Wissmann 1969, 409).[5]

Women were largely excluded from this experience, as evidenced in statements by scientists and in fictional literature. In Adalbert Stifter's novel *The Condor* published in 1844, the female passenger is physically unable to cope with the journey in the balloon. Her male companions' comment on her fainting: "The woman cannot bear the sky".[6] This is

[4] "technisches Vehikel."

[5] "Ort von wissenschaftlichen Messungen und Versuchen, die eine Funktion im Rahmen einer systematischen Gesamterklärung der Beschaffenheit der Atmosphäre und der Vorgänge in ihr hatten"—"der Ort des Vorstoßes des Menschen in einen für den Menschen unbekannten Bereich, der diesen körperlich und mental herausforderte und an seine existentiellen Grenzen trieb."

[6] "Das Weib erträgt den Himmerl nicht."

significant, since Stifter's protagonist's indisposition interferes with both the conduct of the experiments and balloon journeys as an emancipatory act. A closer look reveals that women *did* have their place in the history of ballooning, though not in the field of hard sciences but in popular entertainment.

The tension—and connection—between the culture of spectacle and the culture of science is staged in the film *Aeronauts*. The set designer Charlotte Hutchings let the balloon depart from the Vauxhall Pleasure Gardens in London, which was a bustling Belle Époque fairground at the time Glashier lived—a spectacular place for all masses. While the fictional, filmic James Glashier is busy preparing his instruments and keeping a sponsor of the expedition happy, Amelia Wren rides into the arena standing on a carriage mimicking heroic poses. Wearing a circus costume, she does a flic flac with a somersault on stage and spontaneously performs a clownesque amorous interaction with the reluctant meteorologist to entertain the audience. She fearlessly climbs the balloon's rigging with artistic dexterity, sprays glitter, confetti and fireworks and even throws her trained dog—equipped with a mini parachute—out of the rising balloon. Amelia Wren is an emotional circus performer and initially stands opposite Glashier as a rational scientist. This is evident—among other things—in the fact that she has a feeling for the tempo of an effective performance but not for scientific research methods on atmospheric conditions altering in time. She embodies a heroic showgirl who replaces the male historic figure of Coxwell. This decision by the filmmakers was highly criticised for its historical inaccuracy by Keith Moore (chief Royal Society librarian, in an interview about the film; Bodkin 2018). The lack of understanding about the role of women in aeronautical science springs from the Royal Society's historical aversion to the spectacular use of balloons (determined by women) in entertainment art. In Glashier's times, the Royal Society was unwilling to accept the mingling of science with entertainment culture although it was omni-present. In 1783, the Royal Society ultimately denied the value of balloons in gaining scientific knowledge and (unlike in France) excluded balloon flights as scientifically useful.

Natasha Adamowsky has shown that the strict division between spectacle and experimental culture is a historical fiction, even though it was repeatedly emphasised by nineteenth-century knowledge communities (2010). Adamowsky interprets balloon flights as technically realised miracles and, at the same time, as signs of a rational worldview. The author attributes a special quality of knowledge to the sensually overwhelming form of the flight spectacle, since the affective wonder of balloon travel helped to found meteorology (Adamowsky 2010, 140). Thus, balloons do not mark the alleged lowlands of popular culture (as imagined by the scientific societies); instead, they form the centre of the iconographic self-description of modernity as a time when knowledge was gained and mediated through sensuous experiences (Adamowsky 2010, 182–183). In this regard, it is crucial to note that there are, in fact, historical role models for the figure of Amelia Wren, even if they are little known today. One of them was Sophie Blanchard, who became the captain of an airship and finally an imperial aeronaut under Napoleon I. She died dramatically in 1819 when her balloon burnt over the Tivoli in Paris due to fireworks she had set off. Courageous female aeronauts like Sophie Blanchard refuted the claim of many male scientists (and literary tropes) that women had no business in aviation because of their weak constitution. Harper's film gives women a new visibility by introducing the figure of Amelia Wren now within the context of scientific research. At the climax of the film, Wren's battle with the frozen balloon makes her become a heroine superior in physical strength to the scientist. In doing so, the film bears witness to women in aeronautic culture and its one-sided historisation.

The filmic Glashier corresponds to the type of science hero inscribed in our cultural memory. One of the best-known science heroes is Isaac Newton, who Glashier quotes several times in the film. Monika Mommertz has noted that "knowledge and heroism seems paradoxical", as rationality, objectivity and precision have little in common with the expected physical strength and resilience of a struggling hero (2018, 3). The "intellectual hero" shares certain characteristics with the hero of battlefields, such as "exceptionality, agonality, a commitment to goals considered positive by admirers" (Mommertz 2018, 4). As with all heroes, he is "a fundamentally constructed character and his dependency on heroizing actors and audiences" is clear (Mommertz 2018, 4). Beyond that, however, the science hero has some special qualities, such as the "absolute devotion to scholarly work" and a "willingness if not eagerness, to risk one's own health or well-being, potentially even one's own social or physical death" for a gain of knowledge (Mommertz 2018, 5). His courage "to defend truth" as well as a clear "intellectual superiority" over other people is marked as extraordinary (Mommertz 2018, 5).

It should be kept in mind that this heroism is largely attributed to white men from hegemonic nations. As representatives of colonial empires—in this case Great Britain—and equipped with the self-image of a conqueror, the film certainly updates nineteenth-century hegemonic thinking (Adey 2010) by toning down Glashier's heroism in two ways: lack of practical expertise and being outperformed by a woman. In many respects, Glashier corresponds to a type of hero that was significantly influenced by the British Polar explorers (like Sir James Franklin or Salomon August Andrée)—e.g. thirst for knowledge, transgression of boundaries, fearlessness in the face of the forces of nature. However, he has no idea how to handle the balloon. His lack of function as captain and later fainting have a de-heroising effect. Thus, at the climax of the plot the agency shifts from the unconscious meteorologist to his female companion Amelia Wren. The latter "completes" the male hero through her physical stamina and dexterity, while at the same time revealing his deficiencies. Wren fearlessly takes on the battle with the ice—much like the Polar explorers claimed to do. Although her skin is damaged by frostbites and her hands are in danger of freezing, she fights against the slippery materiality of the icy balloon and ignores the cold. Wren is also sensitive to the environmental changes heralded by the increasing ice or snowfall, e.g. the imminent tearing of the seams due to the increasing pressure in the balloon. Ice is thereby a means of heroising the female figure and de-heroising the male scientific hero; it is her antagonist but also shows her understanding of nature. The film thus becomes part of a conscious shift in the history of knowledge. It increases the recognition of multiple actors in science history and affirms the decontextualisation of heroic narratives.

However, Wren's emancipatory function is countered by moments in which she is characterised as an instinctual being. Sentences such as "My instinct is telling me" imply the character's heroic actions stem less from intellectual capacities than from irrational feelings. This is also reflected in Wren's limited access to rational methods of cognition. In the naturalisation of the female character as a bodily being and the rationalisation of the male character as a thinker figure, the film repeatedly fails to fully overcome the gender inequalities embedded in the history of ice research although it is sensitive to the connection of science and entertainment culture.

Ice as a Tool for Knowledge Production and Its Threat

The clearest interweaving of entertainment and knowledge production of the *Belle Époque* can be seen in James Glashier's historic travelogue *Travels in the Air* from 1871. Glashier was not only a pioneer of meteorology and snow research, but also of scientific communication. In his book, the author describes challenging experiences of extreme cold and his amazement at the beauty of the world in the icy layers of the atmosphere. He combines these experiences with scientific descriptions of scientific methods and data. This conglomerate of factual knowledge with sensual phenomena became largely responsible for the success of his book. However, as a member of the Royal Society, Glashier was of the opinion that ballooning should solely be a scientific endeavour and not phenomena of entertainment culture, meaning that scientists should not stage themselves as adventurers (Glashier 2019, 35). This claim leads to certain contradictions in Glashier's self-description of working with aesthetics to mediate his scientific results. What is more, in his travelogue, Glashier describes the balloon (*Mammoth*) as a wild creature which "struggles impatiently to be free" when challenging its passengers during takeoff (Glashier 2019, 4). In comparison, the film stages ice—in addition to the uncontrollable balloon—as an uncanny and central element within the scientific expedition. Glashier's instruments measure that the temperature dropped suddenly and dramatically after having reached 22,500 ft; at 23,900 ft, the first alarming physical effects appear due to the lack of oxygen in the cold, dry air. The blurriness of the cinematic images mimics the protagonists' declining vision. The resulting identificatory potential gives the viewers the impression of being directly involved in the exposure to the cold and its influence on the human organism. This impression is underlined by the threatening sound of cracking ice on the surface of the balloon. Shortly after Glashier had commented on the increasing hostility of the icy environment—"These are freezing ranges that no one would have predicted"—hypoxia sets in: His mind becomes confused, his sense of sight fails and he finally passes out.

While the film emphasises Glashier's heroic ambitions far more than Glashier's historical persona as author—critical of adventure-seekers—did, it sticks closely to the original travel descriptions of the dangerous effects of extreme cold on the human body. Both the film and Glashier's original travelogue stage ice as a violent force; at the same time is the object of knowledge production. Paradoxically, ice also hinders knowledge of atmospheric conditions. When the environment becomes extreme, "the temperature had fallen to the freezing point [...] the mercury of Daniell's hygrometer fell below the limits of scale" (Glashier 2019, 66). At this point, the liquid inside the instruments freezes and thus neglects to fulfil the equipments' scientific purpose: "I found the water in the vessel supplying the wet-bulb thermometer one solid mass of ice, though I had, by frequent disturbance, kept it from freezing. It did not melt until we had been on the ground some time" observes the historic Glashier (Glashier 2019, 71). Taking this into account, the reader can imagine the dangerous effects of the extreme cold on water within human bodies.

Using the example of the measurement data and research reports of Polar explorers of the late nineteenth century, Benjamin Morgan (2016, 2) pays close attention to the bodily reactions to "extreme cold" as described in researcher travelogues. To the author, these served—and continue to serve—a strategy of verification. Reactions to extreme environments also show that writing and communicating scientific results are closely bound to bodily experiences. As Morgan notes, the "entanglements of human action with geological climatological events" became a "motive force of history" (Morgan 2016, 2) in the period of the Polar conquests and provoked an early ecological consciousness closely linked to physically extreme experiences. The interweaving of factual measurement data with bodily reactions—as already observed in Glashier's notes—evokes a relation between what is felt by the body and what the measuring instruments register. It highlights the entanglement of humans with their environment and their dependence on it. In addition, when examined closely, these notes make clear that the human body always has a role in the exploration of nature.

Hannah Zindel interprets Glashier's diagram "Path of the balloon in its ascent from Wolverhampton to Cold Weston", in his historical accounts as a medium of threshold conditions: "Questions about the spatial boundaries between oxygen-rich and oxygen-poor regions, epistemic boundaries between clear and murky sides of knowledge production, and aesthetic boundaries between representability and unrepresentability can be fixed on this lithograph" (2020, 27). The diagram bears witness to the limits of scientific knowledge production (Fig. 2). It is not a mere coordinate system, but a fusion of a numerical and a non-numerical system. The x- and y-axes in the illustration of the book show the passage of time and the metres of altitude reached in the form of a curve. In contrast, the small

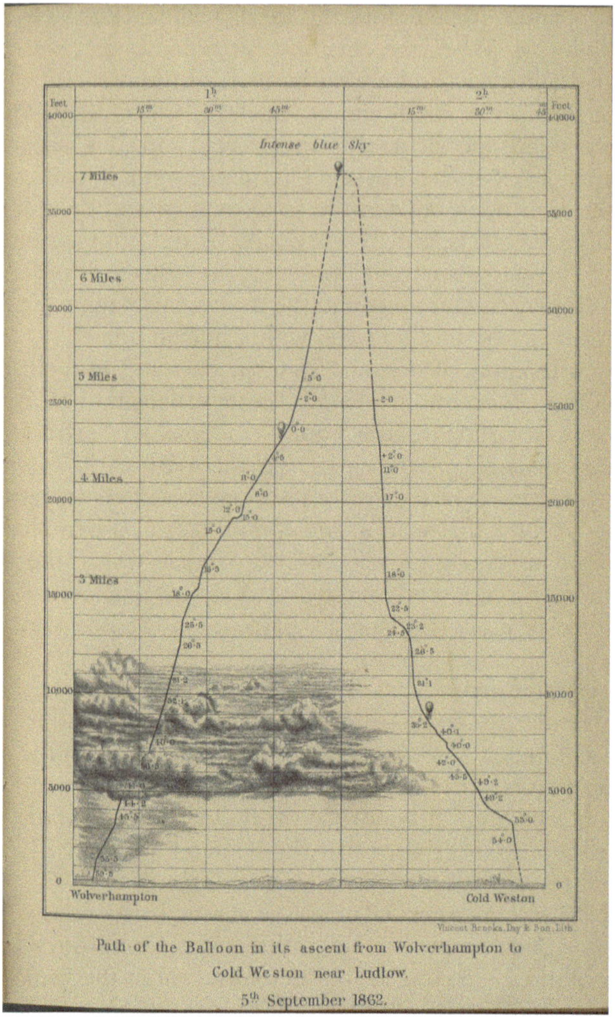

Fig. 2 James Glashier, path of the balloon in its ascent from Wolverhampton to Cold Weston, 5 September 1852, © Staatsbibliothek zu Berlin-Kartenabteilung (Kart. GfE B 3904)

balloon and the landscape stretching out at the bottom (as well as the plastic sea of clouds) are fictional elements which lead to a combination of factual knowledge with imaginary travel experiences. In its plasticity, the balloon is "less an instrument for exploring the vertical than a vehicle for travelling imaginary spaces" (Zindel 2020, 39).[7] Particularly, the upper curve within the scientific diagram marks a transition point from scientific knowledge to a more scientific experienced version of knowledge. During the period in which Glashier was unconscious, the curve turns into a dashed line, which visually signifies the lack of both scientific data and bodily experiences collected during this time (Zindel 2020, 38). Thus, the dotted line communicates how, within the moment of a pioneering scientific deed, "the scene of meteorological measurement (turns) into a platform of physiological introspection" (Zindel 2020, 39).

The film draws on this culture of visualisation of scientific data and feelings of physical exposure by transposing the diagram over sublime cloudscapes throughout the film (Fig. 3). Its frequent appearance marks the progress of the scientific undertaking and further reveals that the film time loosely corresponds with the time it takes the audience to watch the film. Thus, on multiple levels, and analogous to the ballooning experience, the film fuses "spatial practices, knowledge construction and affect" into a single entity by interweaving scientific facts and sensory aesthetics (Dorrian and Pousin 2013, 1). The audience becomes physically involved in the scientific excursion. However, the contrast between the diagram and the

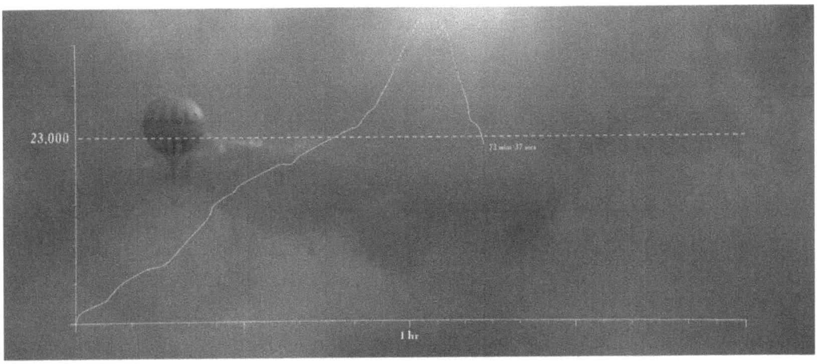

Fig. 3 Harper, Tom, *Aeronauts*, Mandeville Films/Film Nation Entertainment, film still, 2019

[7] "weniger ein Instrument zur Erkundung der Vertikalen als vielmehr ein Fahrzeug, mit dem sich imaginäre Räume bereisen lassen", translated by the author.

vast landscapes also evokes a tilting of perspective between an immersive view into the panorama and a micro-focus lens on the scientific data of the diagram, thus creating a distance between "the visible and the invisible" (Neumann 2006, 356).[8] In essence, by linking the visual culture of scientific research with sublime aesthetics of the Romantic era, the film not only invites an imaginative participation in the most spectacular balloon flight of the Victorian age. It opens up a space of reflection about the scientific history of meteorology; one in which the acquisition of atmospheric knowledge consists of communicated data of measurements, its subjective experiences and interpretations.

Physical Witnessing and Ice Phenomenology

Balloons rarely remained hidden from public perception. The historical Glashier systematically provided testimony to his balloon journeys from various ground points (Tucker 1996, 156). In the *Daily Telegraph*, he advertised a call to the so-called Gentlemen of Science at each launch (there were no calls to any "Ladies of Science"); these mostly ambitious amateurs used instruments to determine the time since the balloon's departure, its altitude and its distance from the ground. Glashier's meteorological network served "to establish and coordinate a system of assimilating empirical contributions from many scattered observers" (Tucker 1996, 156). These gentlemen of science were also part of Glashier's "knowledge production practices" through their reports in the media (Tucker 1996, 160). They made scientific findings public and visible and thus promoted a popularisation of meteorology. Glashier himself often presented his results to a lay audience after his flights; he published in popular journals and even gave lectures at schools, because he was so keen to communicate his scientific findings widely.

It is crucial to note that according to Glashier's historical account, his companion only "nearly" lost consciousness (70), meaning that he was still able to report on the dramatic events. In contrast the film shows a period of time in which both protagonists are unconscious, which makes the viewers the only witnesses to Wren's heroic deed (opening the valve and near-death slide off the frozen balloon) and the breaking of the world record (Fig. 4). When the female protagonist climbs up the icy dome of the balloon—visualised in the film with an "on top of the world aesthetic"—there

[8] "Grenze zwischen dem Sichtbaren und dem Unsichtbaren", translated by the author.

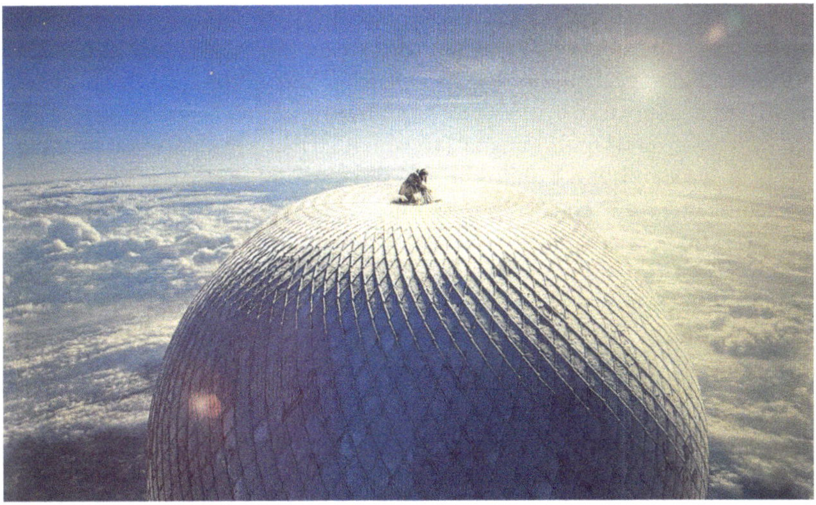

Fig. 4 Harper, Tom, *Aeronauts*, Mandeville Films/Film Nation Entertainment, film still, 2019

are no witnesses except the spectators. A large audience, on the other hand, was present when the filmic balloon took its spectacular departure from the fairground. As in the archival documents of the travel, the witnessing of the journey in the balloon basket and from the ground through telescopes is central to the film. The film shows Glashier in his younger years, as he viewed the flight of a balloonist through a telescope. During his own journey, members of the Royal Society take over this task. (Scientific) Heroes and heroines are dependent on their perception and medial mediation, i.e. heroes and heroines can only be constituted and affirmed by an audience (Asch and Butter 2016, 13). In the film, there are not immediate but only delayed “hero-makers” for the altitude record, such as *The Times* newspaper writing (read aloud by one of the movie figures) “Two aeronauts have been nearer to the moon and stars than all the race of man before them”. However, there are always moments of identification and participation in the film that sensually appeal to viewers as witnesses, especially in the sensuous effect of cold, the affecting glitter of the icy surface of the balloon and the ominous sounds of cracking ice.

Thus, the film audience is framed as part of a scientific community and witness to its yields; this is evoked verbally when the film character and historical Glashier marvel at the overwhelming grandeur of the sky panoramas and the beauty of the weather phenomena. Glashier's historical persona made the balloon both a scientific and a "philosophical instrument" that generated worldly, and self-awareness through beauty of different atmospheric phenomena (Glashier 2019, 35). Since Plato, flying has been a metaphor for philosophical thinking in the sense of questioning the familiar (Trautsch 2020, 24). In Merleau-Ponty's phenomenology, however, flight was explicitly rejected as a metaphor for thinking, since the purely visual experience of the world lacked the haptic sense (Merleau-Ponty 1995, 59; Trautsch, 33). This reservation seems to apply more to aeroplane travel, whereas floating in an open basket below a gas balloon is a thoroughly physical experience; especially when the balloon ascends to extremely icy heights. In fact, balloon journeys can, increasingly, be described as "sensuous mobilities" in the air (Adey 2010, 19), since travel accounts include not only what is seen (ice) but also what is felt (piercing cold).

It is crucial to note, by the way, that the film differentiates between snow and ice. Snow has a narrative and relieving function: It appears when it comes to moments of decision, for example when Wren encounters Glashier's book on snowflakes. The snowfall after the narrowly averted catastrophe is perceived as a liberation by both protagonists, for it signals ice melt. Snow also falls when Glashier reveals to Wren the real driving force behind his artistic drawings: "(Science) helps to give meaning to the many things we can't control. It brings a degree of order to the chaos that surrounds us. But whilst we may be able to explain the science behind an aureole or the falling snow it's not possible to account for its beauty." In the film, snow is a symbol for the sublime beauty of nature, which, because of its ultimate mysteriousness, is recognised as something different to humans and, as such, is to be respected and admired.

In summary, insights into the physical entanglement of humans with nature, humans' dependence on nature and its significance as a powerful entity are expressed and generated in the film through the use of snow and ice. Ice functions as an active agent in the history of science; it is a medium of knowledge-making, a tool of aesthetic science communication and an element of philosophical reflections. In this sense, the film *Aeronauts* is not only a biopic of the historical figure of James Glashier, or a revealing movie about the ignorances and driving forces of science history. *Aeronauts* is a historic and current portrayal of ice as an evidence, threat and agent. It features ice as a scientific and aesthetic device. Ice refers to the invisible interconnections of the planet, thus drawing a line between former understandings of nature and today's climate catastrophes. Hence, humans' attitude and responsibility towards nature is questioned in light of its ever-present powerful roots in the era of heroic science research and current ecological debates.

Bibliography

Adamowsky, Natascha. 2010. *Das Wunder in der Moderne: Eine andere Kulturgeschichte des Fliegens*. München/Paderborn: Fink Verlag.

Adey, Peter. 2010. *Aerial Life: Spaces, Mobilities, Affects*. Chichester: Wiley-Blackwell. https://doi.org/10.3167/TRANS.2012.0202013.

Asch, Ronald, and Michael Butter. 2016. *Bewunderer, Verehrer, Zuschauer: Die Helden und ihr Publikum*. Würzburg: Ergon Verlag.

Bloom, Lisa. 2022. *Climate Change and the Polar Aesthetics: Artists Reimagine the Arctic and Antarctic*. Durham/London: Duke Press.

Bodkin, Henry. 2018. Ballooning Hero 'Airbrushed' from History to Make Way for Female Character in Eddie Redmayne Film. *The Telegraph*. https://www.telegraph.co.uk/news/2018/08/15/ballooning-hero-airbrushed-history-reunite-eddie-redmayne-felicity/. Accessed 19 Aug 2022.

Boetzkes, Amanda. 2010. *The Ethics of Earth Art*. Minneapolis/London: University of Minnesota Press. https://doi.org/10.5749/minnesota/9780816665884.001.0001.

Brady, Emily. 2013. *The Sublime in Modern Philosophy*. Cambridge: University Press. https://doi.org/10.1017/CBO9781139018098.

Chichester, K. Lee. 2021. "Snowflake Generation": Die Kristallisierung kosmischer (Un-)Ordnung. In *Kältebilder: Ästhetik und Erkenntnis am Gefrierpunkt*, ed. Matthias Bruhn, 10–27. Berlin/Boston: De Gruyter. https://doi.org/10.1515/9783110747614.

Dorrian, Mark, and Frédéric Pousin. 2013. Introduction. In *Seeing from Above. The Aerial View in Visual Culture*, ed. Mark Dorrian and Frédéric Pousin, 1–10. London/New York: I. B. Tauris.

Elias, Amy, and Christian Moraru. 2015. *The Planetary Turn. Relationality and Geoaesthetics in the Twenty-First Century*. Evanston: Northwestern University Press. https://doi.org/10.2307/j.ctv3znz1s.

Gamper, Michael. 2011. Der Ballon als multifunktionale Versuchsanstalt. Stifters "Der Condor" als erweitertes Experimentalsystem. In *Magie der Geschichten. Weltverkehr, Literatur und Anthropologie in der zweiten Hälfte des 19. Jahrhunderts*, ed. Kerstin Stüssel and Michael Neumann, 404–416. Konstanz: University Press. https://doi.org/10.1515/jdrg-2013-0014.

Glashier, James. 2019. *The Aeronauts. Travels in the Air.* Brooklyn/London: Melville House.

Gründler, Hanna. 2019. *Die Dunkelheit der Episteme. Die Kunst des aufmerksamen Sehens.* Berlin: Reimer Verlag.

Harper, Tom. 2019. *Aeronauts*, Mandeville Films/Film Nation Entertainment, 101 minutes.

Hemkendreis, Anne. 2021. Icy Hieroglyphics: Wilson Bentley's Snow Crystals. *w/k: Between Science and Art*, 15 November. https://doi.org/10.55597/e7077.

Horn, Eva, and Hannes Bergthaller. 2019. *Anthropozän: Zur Einführung.* Hamburg: Junius Verlag.

Merleau-Ponty, Maurice. 1995. In *Das Sichtbare und das Unsichtbare*, ed. Claude Leford. München/Paderborn: Fink Verlag.

Mommertz, Monika. 2018. Heroization in Science, Scholarship and Knowledge-Production: The intellectual Hero in Transdisciplinary and Trans-epochal Perspective. *helden.heroes.héros* SI (4): 3–12. https://doi.org/10.6094/helden.heroes.heros./2018/HS/01.

Morgan, Benjamin. 2016. After the Arctic Sublime. *New Literary History* 47: 1–26. https://doi.org/10.1353/nlh.2016.0000.

Neumann, Gerhard. 2006. Fernrohr, Mikroskop und Luftballon. Wahrnehmungstechniken und Literatur in der Goethezeit. In *Spektakuläre Experimente. Praktiken der Evidenzproduktion im 17. Jahrhundert*, ed. Helmar Schramm, Ludger Schwarte, and Jan Lazardzig, 245–377. Berlin/New York: De Gruyter.

Schneider, Birgit. 2021. Sublime Aesthetics in the Era of Climate Crisis? A Critique. In *The Routledge Companion to Contemporary Art, Visual Culture and Climate Change*, ed. T.J. Demos, Emily E. Scott, and Subhankar Banerjee, 263–273. New York: Routledge. https://doi.org/10.4324/9780429321108.

Trautsch, Asmus. 2020. Bodenlos sicher: Phänomenologie der Flugreisen. In *Die Phänomenologie der Flugreise. Wahrnehmung und Darstellung des Fliegens in Literatur, Film, Philosophie und Populärkultur*, ed. Jan Röhnert, 20–44. Köln: Böhlau Verlag. https://doi.org/10.7788/9783412500863.front.

Tucker, Jennifer. 1996. Voyages of Discovery on Oceans of Air. Scientific Observation and the Image of Science in an Age of "Balloonacy". *Osiris* 11: 144–176.

Wissmann, Gerhard. 1969. *Geschichte der Luftfahrt von Ikarus bis zur Gegenwart. Eine Darstellung der Entwicklung des Fluggedankens und der Luftfahrttechnik.* Berlin: VEB Verlag.

Zindel, Hannah. 2020. *Ballons: Medien und Techniken früher Luftfahrten.* München/Paderborn: Fink Verlag. https://doi.org/10.30965/9783846764510.

Hard Ice, Soft Snow? Transnationalism, Spectatorship and the Arctic Sublime in *Chasing Ice* (2011) and *Silent Snow* (2011)

Johannes Riquet

In what is probably the best-known scene of the award-winning American climate change documentary *Chasing Ice* (2011), directed by Jeff Orlowski, we see a glacier on Greenland's west coast (Sermeq Kujalleq) calving into the Ilulissat Icefjord. The film follows nature photographer James Balog's mega-project of installing cameras throughout the Arctic, mostly in Greenland, to monitor and make visible the effects of climate change over several years. At the beginning of the calving scene, we see Balog and one of his team members with their cameras, communicating with each other by phone. As the glacier starts to calve, the initial comments soon cease and for almost two minutes we only see and hear the spectacular images and sounds of enormous masses of ice breaking into the sea. This is followed by footage of Balog lecturing on the calving

J. Riquet (✉)
Tampere University, Tampere, Finland
e-mail: johannes.riquet@tuni.fi

A. Hemkendreis, A.-S. Jürgens (eds.), *Communicating Ice through Popular Art and Aesthetics*, Palgrave Studies in Media and Environmental Communication,
https://doi.org/10.1007/978-3-031-39787-5_11

event: "That's a magical, miraculous, horrible, scary thing. I don't know that anybody's really seen the miracle and horror of that."

For anyone familiar with eighteenth- and nineteenth-century aesthetic theory, Balog's words evoke the sublime, whose characteristic mixture of terror and awe is predicated on a separation of the human observer from the overwhelming shapes and forces of the natural world. While the best-known definitions of the sublime—those of Edmund Burke and Immanuel Kant—differ in key respects, "each tells a story in which some form of distance from danger provides the necessary condition for converting painful sensations into pleasurable contemplation" (Morgan 2016, 14). *Chasing Ice* indeed distances the observers from the spectacle of ice, creating a separation not only through the camera but also through the seamless move from the ice fjord to the lecture hall. The sublime presentation of the glacier's calving in *Chasing Ice* thus generates a double paradox. Indexical of the terror of climate change, it casts Balog's team as witnesses to an overpowering, sublime force—even while Arctic ice is presented as threaten*ed*, rather than threaten*ing* as per the nineteenth-century tradition (see Loomis 1977). Furthermore, Balog's characterisation of the calving event as both "miracle and horror" presents it as frightening *and* enjoyable. The film's economy of desire, in other words, (partly) counteracts its own rhetoric of climate activism: it is predicated on a pleasurable but passive retreat into vision.

Chasing Ice thus reinvents the Arctic sublime by presenting climate change as a visual spectacle recorded by a photographer—who becomes a twenty-first-century version of the heroic Arctic explorer—in the interest of teaching global audiences about the extent and speed of climate change. The sublime is here elevated to the level of the planetary, leaving the (nineteenth-century) realm of the national (Bloom 1993; Hill 2008). As such, *Chasing Ice* is an example of an "environmentalist documentary film" (Truscello 2018), a genre that mobilises documentary's potential for social and political critique to support climate activism and raise ecological awareness. Filmmakers use documentary to illustrate the unimaginable scale and temporality of the climate crisis (Hilderbrand 2020, 211) and to visualise scientific data (Kaganovsky et al. 2019b, 21). Environmental documentary film and climate science, in other words, are entangled in multiple ways, with the Arctic playing an important role in both: "In the Anthropocene, documentary moving images are the primary mode through which climate change is conveyed, as the Arctic is one of the regions most affected by the climate crisis" (Kaganovsky et al. 2019b, 21).

As Lilya Kaganovsky, Scott MacKenzie and Anna Westerstahl Stenport (2019b) argue, the Arctic plays a central role in the history of documentary film; indeed, Robert Flaherty's *Nanook of the North* (1922) was championed by John Grierson as one of the foundational works of the genre (Barsam 1992, 79). If the history of the documentary is closely linked to the Arctic ethnographic film and Northward imperialist expansion, then "in the age of climate change, the Arctic has again taken center stage in the realm of moving images" (Kaganovsky et al. 2019b, 22). Once again, this interest coincides with an international scramble for the Arctic and its resources (Dodds and Nuttall 2016).

If *Chasing Ice* is representative of a type of film that reinvents aesthetic and epistemological traditions inherited from colonial and imperial imaginaries, other documentaries have developed what Lisa Bloom has termed a "new polar aesthetics" (2022). For Bloom, these films "challenge[...] the dominant narrative of mainstream media, which equates climate change with apocalyptic spectacles of melting ice and desperate polar bears, and green capitalism with masculinist imagery of sublime wilderness and imperial heroics" (2022, 2). A good example of the latter type of films, though not mentioned by Bloom, is *Silent Snow* (2011) by Greenlandic filmmaker Pipaluk Knudsen-Ostermann and Dutch director Jan van den Berg, which intersperses two journeys—one in Greenland and one around the world—to trace the pollution of the Arctic through the global circulation of pesticides. This chapter offers a comparative analysis of the two films, focusing on their different treatments of ice and snow. While *Chasing Ice* posits a universal human subject that aligns with certain forms of Anthropocene discourse, *Silent Snow* ties together art and activism in a transnational aesthetics that is grounded in relational experiences and Inuit understandings of snow and ice, contrasting with the masculinist, individualist connotations of ice in Orlowski's film. While both films are invested in communicating and visualising environmental issues, and engage with snow and ice materially as well as figuratively, they do so in radically different ways that also posit very different viewers: a passive consumer in *Chasing Ice*, and an active participant in *Silent Snow*.

Reinventing the Arctic Sublime in the Twenty-First Century

In his 1757 treatise *A Philosophical Enquiry into the Origin of our Ideas of the Sublime and Beautiful*, Edmund Burke famously defined the sublime as follows: "The passion caused by the great and sublime in nature, when those causes operate most powerfully, is astonishment; and astonishment is that state of the soul, in which all its motions are suspended, with some degree of horror" (1990, 53). While eighteenth-century discussions of the sublime typically focused on awe-inspiring experiences of the rugged mountainous and glacial landscapes of the Alps (or, at a smaller scale, the crags of the English Lake District), the Arctic craze of nineteenth-century exploration culture turned the polar regions into a paradigmatic space of sublime experience. In his seminal article on the Arctic sublime, Chauncey C. Loomis writes:

> Edmund Burke, had he written his analysis of [...] the sublime a century later [...], might well have used the Arctic in his discussions of light and dark, sound and silence, obscurity, solitude, vastness, and magnificence as sources of sublime astonishment and terror. (1977, 101–02)

The sublime terror of Arctic landscapes had a prominent place in nineteenth-century British and American visual culture, from landscape paintings to panoramas and magic lantern slides. Thus, Russell A. Potter outlines "parallel histories of exploration and exhibition", arguing that the pervasive presence of the Arctic as a visual spectacle in metropolitan centres was rooted in "a deep cultural and geographical cathexis between new technologies of vision and the regions of the earth most difficult—and terrifying—to behold" (2007, 12).

In this context, the Arctic sublime is "often understood as an ideologically significant aesthetic category: [...] it generated triumphalist narratives about national character, masculinity, or geopolitics; and it often caricatured or erased the indigenous people of the Arctic" (Morgan 2016, 3). As Jen Hill argues in relation to John Franklin's *Journey*, for instance, "readers' encounters with the sublime [...] enabled them to participate in the expedition's overcoming of Arctic threat, and thus to read Franklin's text as evidence of a British male national character uniquely suited to imperial mapping projects" (2008, 44).

Without explicitly invoking the sublime, Bloom similarly draws attention to the ideological dimensions of aesthetics in American polar exploration, arguing that the Arctic "literalized the colonial fantasy of a tabula rasa where people, history and culture vanish. The absence of land, peoples, or wildlife to conquer [...] allowed the discovery of the North Pole to appear above political and commercial concerns" (1993, 2–3). As these analyses make clear, the Arctic—notably the ice- and snow-covered Arctic—functioned as a gendered stage on which masculinist imperial fantasies were projected and enacted.

The sublime continues to exist in contemporary Arctic imaginaries, as Bloom notes (2022, 5). This continuity is perhaps most evident in the conjunction of science, politics and exploration that characterises nineteenth-century polar voyages (Renov 2019, 209), Cold War expeditions to the Arctic (Farish 2006) and contemporary climate science. These waves of interest in the Arctic go hand in hand with a "documentary desire" (Renov 2019, 208–213) for the polar regions that has conjoined technology and aesthetics since the nineteenth century. Polar explorers in the nineteenth century routinely took visual artists on their voyages (Renov 2019, 209) and "[s]hortly after the invention of photographic technology, cameras and photographers were regularly included on expeditions to the Arctic" (Belanger 2019, 220). The Arctic sublime extended to the level of the medium itself, as it was notoriously difficult to capture polar landscapes on camera because of the equipment's sensitivity to cold and other environmental challenges (Belanger 2019, 220). The titular chasing of ice in Orlowski's film thus refers not only to the disappearance of sublime landscapes caused by climate change but also to a technological quest; to the photographer's (and filmmaker's) desire to capture an Arctic spectacle of snow and ice. While the first of these meanings is clearly committed to climate activism—even while it reinscribes fatal impact narratives of a doomed Arctic—the second meaning suggests the consumption of the Arctic through commodified images, unwittingly repeating the capitalist quest for (Arctic) resources that is responsible for the current climate crisis in the first place.[1]

[1] Truscello makes a similar point when he characterises *Chasing Ice* and related films as "eco-opportunistic", arguing that "the word capitalism must not be uttered in the film, and the solution to the ecological crises created by capitalism must also involve capitalism" (2018, 270). In Truscello's view, however, *Chasing Ice* "understate[s] the crisis" while it succeeds in eschewing the pitfalls of catastrophism; in contrast, I argue that the film's catastrophist rhetoric is in fact a key symptom of its complicity in the structural logic of capitalism.

The film's implication in an economy of desire linked to the global circulation of images in late capitalism raises interesting questions about the changed status of the Arctic sublime in the twenty-first century. As Benjamin Morgan argues,

> the story that the Arctic now tells us seems to have less to do with ideologies of the nation-state, and more to do with the ways in which extranational networks of humans and nonhumans distributed across the entire planet will be affected, unequally, by the changes that are manifested most evidently at the earth's ice caps. (2016, 3)

The new Arctic sublime of climate change, too, operates in an extra- or supranational way as it points to the largely invisible forces of planetary ecosystems. As such, the sublime is no longer tied to the cultural imaginary of individual nations, but rather to the fate of the planet. The sublime thus remains linked to the ungraspable and inaccessible, but the feelings of terror and awe now respond to human-induced effects on the environment as much as to the environment's impact on humans. This also, however, potentially confuses agencies and affects, leading to a medially generated mixture of terror and awe that distracts from the structural issues and inequalities that underlie climate change.

Chasing Ice: Masculinity, Spectacle and the Photographer-Explorer

The changed status of the Arctic sublime in the Anthropocene impacts our reading of the scopophilic regime in *Chasing Ice*, as a detailed analysis of its cinematic and narrative properties makes clear. As suggested above, the characteristically sublime experience of fascination, awe and fear in the film becomes a medial event that gives the viewer access to something that seemingly *nobody has ever seen before*. The scene discussed in the introduction accentuates this by creating an identification between internal camera, film camera and viewer: before the calving begins, we see one of the cameras from behind in an approximation of an over-the-shoulder shot, with the camera in the place of a human subject (Fig. 1). This arrangement repeats the classic composition in Romantic paintings, such as Caspar David Friedrich's *Wanderer above the Sea of Fog* (1818), where we see an observer gazing into a sublime landscape from behind, functioning as a surrogate for the spectator. *Chasing Ice* replaces the human gaze of the

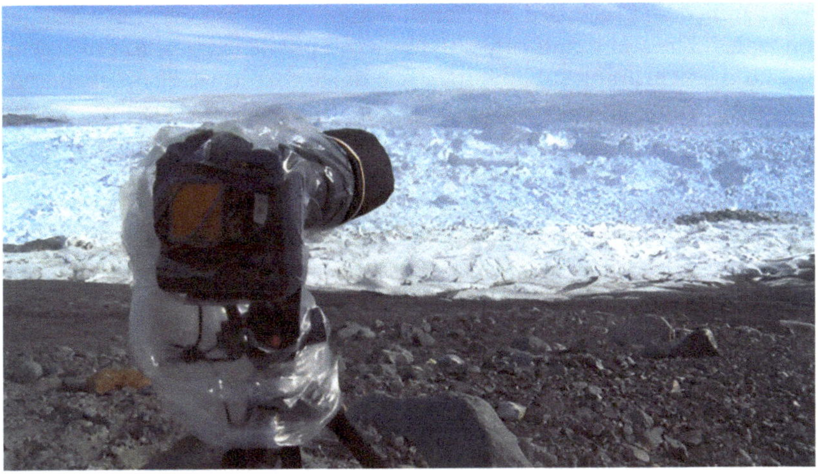

Fig. 1 The technologically mediated gaze: film still from Jeff Orlowski, *Chasing Ice* (2011)

Romantic and Victorian sublime with a technological gaze, which is then transferred to the viewer: after the initial shots of the technological apparatus, we no longer see any internal cameras. Their place merges with that of the film camera, and thereby with our own.

The position created for the viewer corresponds to that stipulated by Jean-Louis Baudry in his psychoanalytic theory of the cinematic apparatus.[2] For Baudry, the viewer is constituted as a subject by taking the position of the camera, experiencing an illusion of completeness by functioning as a point of convergence for, and providing imaginary coherence to, the world of the film. The screen thereby functions like the mirror in Lacanian theory. What is crucial for Baudry, however, is the link between this identification and ideology: "The ideological mechanism at work in the cinema seems thus to be concentrated in the relationship between the camera and the subject" (1974–1975, 46). In *Chasing Ice*, the viewing subject is placed in a position that conjoins visual pleasure and passivity. At the end of the scene, the cameramen make this explicit in a voice-over comment: "We're just observers. These two little dots on the side of the mountain.

[2] I thank Tatu Laukkanen for suggesting this to me after my lecture on *Chasing Ice* at the master class "Media and the Arctic", University of Tampere, 27 November 2018.

We watched and recorded the largest witnessed calving event ever caught on tape." The scene thus offers the viewer a privileged position of accessing an overwhelming sublime spectacle of ice that leaves them as speechless as the cameramen. Visual spectacle and helplessness thereby go hand in hand. This results in a contradiction in a film whose declared intention is to raise awareness of climate change and incite action. Overall, the scene creates the impression of conveying a sensation that has not been captured on camera before; the most important thing here and throughout the film seems to be the aesthetic spectacle, and the sense of being first in capturing it. We do not necessarily get a sense of these images signifying a warming earth—instead, we see an overwhelming spectacle of ice. Officially, the film tells us that the Arctic landscapes are threatened by human action but visually, this is inverted: as in the earlier accounts of Arctic explorers, it is the ice that appears to threaten human life.

If "cinema can thus appear as a sort of psychic apparatus of substitution, corresponding to the model defined by the dominant ideology" (Baudry 1974–1975, 46), *Chasing Ice* presents the figure of the photographer and filmmaker as a modern-day explorer in a new sublime landscape of technologically mediated climate change spectacle.[3] Indeed, the film repeatedly celebrates the photographer-explorer's heroic daring in obtaining these images, and the occasional failure of the equipment only underscores this. In one of the first shots of the film, and then again at its end, we see Balog on the edge of the icy sea, struggling against the elements, camera in hand: a hyper-masculine individual in a desolate landscape. In many scenes, the film effects a figurative transfer from the hardness of the ice to the photographer's hard masculine body. It emphasises his heroic, selfless perseverance despite several knee operations, and the camera repeatedly dwells on his bruised, suffering body; in one scene, Balog talks about the crunching in his knee while we hear the ice crunching under his feet. Here, the powerful strength and spectacle of the ice infuse the twenty-first-century explorer. In fact, yet another transfer takes place in the second part of the film, when the spectacle is no longer the ice, but the photographer-explorer himself: the film stages him as a celebrity and star seen on multiple screens and in front of large audiences, who gaze at him in silent, and sometimes open-mouthed, awe, the same way they gaze at his spectacular footage.

[3] Indeed, Kaganovsky, MacKenzie and Westerstahl Stenport mention *Chasing Ice* as an example of "'new explorer' films that draw on past tropes" (2019, 6).

The ideological gesture of the film that turns the photographer into a modern-day Arctic explorer also reinscribes colonial fantasies. Like the sublime tradition it adapts, *Chasing Ice* presents the Arctic as empty of humans; throughout the film we do not see a single Indigenous inhabitant of the Arctic. Even if the film manages to make climate change visible in impressive—and no doubt educational—ways,[4] the subject position it creates for the spectator is inherited from the colonial sublime. On the one hand, climate change is figured as an aesthetic spectacle of an inevitable threat that reduces humans to passivity and helplessness, rather than suggesting possibilities for agency, response and responsibility. On the other hand, it is specifically the Arctic as a lived environment that is disavowed by the film; as in the discourse of salvage anthropology, the medial record matters more than the 'real thing', but here the humans who live in the Arctic do not even have a *medial* role to play. As Westerstahl Stenport and Richard S. Vachula argue, the Western (or 'Southern') construction of the Arctic as an empty space has made it hard to think of it as entangled with our everyday lives: "The marginalization of indigenous peoples of the Arctic and the characterization of the region as otherworldly has effectively dissociated the Arctic region and the climate change associated with it from the sphere of the Western society" (2017, 293). The tradition of the Arctic sublime has thus constructed the Arctic as a separate sphere—a spectacular, threatening and aestheticised realm dissociated from the ordinary lifeworld of the Western subject. Contemporary climate change discourse sometimes taps into this tradition; one only needs to think of Greenpeace's Save the Arctic campaign, which uses polar bears and icebergs to create empathy but disregards the human inhabitants of the Arctic.

Chasing Ice is thus subject to Timothy Morton's critique of the "ecologocentrism" that underlies much of Western environmentalism in affecting a dissociation between the (Western) subject and an aestheticised natural world (2007). The narcissism characterising Western climate science is also critiqued by the Greenlandic politician Aqqaluk Lynge, former president of the Inuit Circumpolar Council:

[4] Thus, Klaus Dodds and Rikke Bjerg Jensen write that "the film picks up on [the] trope of the ice being indicative of rapid change" in the hope of convincing audiences of "the severity of ongoing climate change" (2019, 157), and for Michael Renov, the film "becomes a powerful tool for altering the public conversation around climate change" (2019, 211).

> With all the flurry of scientific enquiry on this issue, one could easily be led to believe that it is the researchers who are the most affected by the world's changing climate, and not the Inuit. I plead with western scientists to be careful how you conduct your research in our land. Work with us as equal partners and not as the colonizers and missionaries did. Help us deal with not only your own interesting research, but with our concerns such as how to deal with industry, which is keen to see an Arctic sea route open up to *them*. (2009; emphasis in original)

Lynge's critique pertains to the discursive frames of climate change as much as to the actual scientific activities conducted, both of which are targeted in Igloolik Isuma Productions' film *Qapirangajuq: Inuit Knowledge and Climate Change* (2010). Knudsen-Ostermann and van den Berg's *Silent Snow* does not directly target Western climate science—indeed, it is not directly about climate change at all—yet its form of communicating environmental problems is radically different from that of *Chasing Ice* and implicitly critiques the top-down environmentalism of Southern discourses about the Arctic.

The Poetics and Geopolitics of Snow and Ice

In both films, the cryosphere functions as a medium to communicate environmental issues, but their environmental activism is grounded in very different understandings of snow and ice. Therefore, before turning to the analysis of *Silent Snow*, I will reflect on the cultural and geopolitical significance of how snow and ice are imagined. In his recent book *Ice: Nature and Culture* (2018), Klaus Dodds writes that "[i]ce is integral to the human condition" (2018, 23). While ephemeral, ice has shaped—and continues to shape—landscapes, regulates the global circulation of water and climate, sets limits to and enables transport and travel, structures and stabilises the ground in many parts of the world, stores and preserves matter and information, and functions as an archive of the earth's history (cf. Dodds 2018). The physical properties of ice give it a special status as solid yet at the same time mobile, fluid and unpredictable matter. Ice embodies the dynamics of space (cf. Massey 2005). In the Arctic Ocean, for instance, it challenges the distinction between land and sea and frustrates human attempts to draw boundaries, to cartographically and politically fix space (see Steinberg and Kristoffersen 2017, 625). Ice also plays an important role in the realms of language and the imagination; the English language is pervaded with icy and snowy metaphors (see Hansson and Norberg 2009). In literature, film and popular culture, ice fascinates at the same time as it repels and scares; it is the site of cultural anxieties and fantasies. As Dodds points out, "[w]e have an ambivalent relationship with ice and snow" (2018, 9), yet there is no generalisable, universal 'we' that is not culturally inflected: the cultural production of snow and ice varies across spaces, times and cultural configurations.

While the Western tradition of the sublime imagines snow and ice as a sphere of desolation and death—a powerful antagonist of (human) life—Inuit often imagine ice and snow in very different terms. Lynge expresses his respect for ice in an account of a calving event that both resembles and radically differs from the one presented in *Chasing Ice*:

> When you think of Greenland, you likely think of glaciers and icebergs, and more generally, of ice. I hope that after this talk you will also think of my people and our intimate relationship *with* the ice. In the Inuit language, we use the word *sila* for ice. But *sila* also means much more than ice. It also means weather, climate, environment, sky, and indeed, the universe. Although I have never been a fulltime hunter, like all Greenlanders, I grew up around ice and observed its behaviour. […] One must respect, and sometimes fear ice. It is the giver of life for us. Fish are drawn to the nutrient-rich waters at the base of the freshwater icebergs, which in turn, brings seabirds and seals. I was once hunting near a most magnificent iceberg possessing unimaginable colours and peaks and towers rising high into the beautiful blue sky. […] The *opera house* towering above us suddenly began to heave, and groan, and sway. I had never been so dangerously close to a calving iceberg before. (2009)

Lynge's description connects ice to life rather than death, detailing its multiple ecological functions as well as the perceptual richness it generates. As in the sublime tradition, ice is personified, but not as an antagonist; instead, Lynge emphasises the intimacy of the encounter with ice, which is figured as a lifeworld and complex relational environment. Drawing, among other examples, on Lynge's account of ice, Kirsten Hastrup argues that ice in Inuit conceptions is a space that generates stories (2013, 56–60), which is explicit in the preface of the edited collection *SIKU: Knowing Our Ice*:

> The ice transforms not only the physical landscape around us but also the emotional landscape within us. We welcome the return of the sights, sounds,

> and even smells and tastes that come with the sea ice. […] Along with the senses of sea ice come the stories, the journeys, and the memories. […] The sea ice has always been a place for the making and telling of good stories, from the fantastic travels of great shamans to the mythical beings that live on and under sea ice, like the *Qalupaliit* ("people snatchers"). (Sanguya and Gearhead 2010, ix–x)

Like Lynge's story, this account emphasises the sensory richness of ice as well as portraying it as a sphere where different living beings interact and from which stories emerge. In both cases, ice is tied to intimate affects, intense phenomenological experiences and narratives that connect it to humans. Indeed, the beginning of Lynge's story ("I was once hunting …") not only is a narrative strategy of grounding knowledge in lived experience that is common to Inuit epistemologies and storytelling conventions (cf. Martin 2012, 98–120), but also emphasises the hunter's connection to the icy landscape. At the same time, Lynge's account contains elements that could be considered sublime. "[U]nimaginable colours", impossibly tall "peaks and towers" that reach up "into the beautiful blue sky": the overwhelming and monumental power of the ice is here coupled with its indescribable (or "unimaginable") quality. The metaphor of the opera house recalls the comparison of the icebergs to Manhattan's skyscrapers and the staging of a spectacle in *Chasing Ice*, whereby the reference to one of the palaces of Western art suggests that Lynge self-consciously frames his description in relation to Western aesthetic sensibilities and traditions (he is, after all, addressing a non-Indigenous audience).

At the same time, Lynge's account also shows that a reframed version of the sublime is not altogether incompatible with Indigenous perspectives on the Arctic environment. Michael Renov suggests as much in an article on the sublime in Arctic documentary film. While not writing about Indigenous cinema, he is interested in a more contemplative version of the sublime. Drawing on Barry Lopez's *Arctic Dreams*, he suggests that the embodied attentiveness to the land and its life in Inuit hunting culture resonates with the meditative sublime in a number of experimental films about the Arctic. Thus, he quotes Lopez's observation that Inuit hunters "faced nature with *ilira* (nervous awe) and *kappia* (apprehension)" (qtd. in Renov 2019, 208). While Renov does not seriously engage with Inuit aesthetic experience, his passing comments connect with Kaganovsky, MacKenzie and Stenport's suggestion that "northern Indigenous concepts might help articulate alternative and complementary approaches to representations of the Arctic" (2019b, 13); as an example, they mention the Inuit concept of *sila*, which is prominently used by Lynge in the speech cited above. *Sila* plays an important role in Inuit philosophy, with meanings ranging from 'air', 'weather' and 'wind' to 'spirit', 'mind' and even 'all' (Qitsualik 2013, 29). As a concept that denotes human-environmental relations, it gestures towards a form of communicating environmental challenges in the Arctic that is very different from the sublime rhetoric of *Chasing Ice*. As such, it provides a good entry point into the second film discussed in this chapter, the environmental documentary *Silent Snow*.

Silent Snow: Planetary Entanglements and Transnational Networks

Like *Chasing Ice*, *Silent Snow* takes an educational stance in focusing on the Arctic to address global environmental challenges.[5] However, while Orlowski's film presents the Arctic only as a symptom and synecdoche of global climate change—indeed, the effects of global warming on life in the Arctic are invisible in the film—*Silent Snow* reverses this logic and moves from the planet to Arctic life, foregrounding how industrial pollution across the world has led to the poisoning of Arctic fish and sea mammal populations, in turn affecting human bodies through food consumption.[6] The filmmakers are particularly interested in the poisoning of lands and bodies through persistent organic pollutants circulating globally in water and air. The film thereby draws on research by Indigenous and non-Indigenous researchers, such as Sheila Watt-Cloutier's work on how seal and whale blubber with high concentrations of persistent organic pollutants have poisoned the breast milk of Indigenous mothers, "caus[ing] irreversible neurological damage to infants" (Bloom 2022, 91; see also Watt-Cloutier 2002). Indeed, the film takes its title from Marla Cone's 2005 book *Silent Snow: The Slow Poisoning of the Arctic*, whose first chapter, titled "The Arctic Paradox", refers to the fact that the Arctic has been disproportionately affected by global industrial pollution while only

[5] For Kaganovsky, MacKenzie and Stenport, *Silent Snow* is an example of "Arctic Indigenous environmental documentaries" (2019, 23); Lill-Ann Körber similarly discusses the film "[a]s an example of ecocinema from and about Greenland" (2017, 156).

[6] As Körber argues, "[t]he main point of the film is the juxtaposition of local and global perspectives, of a 'sense of place' with a 'sense of the planetary'" (2017, 156).

minimally contributing to it (Cone 2005, 15–110; see also Bloom 2022, 90–91).

The film does not merely explain this planetary entanglement, but also performs it in narrative and aesthetic terms. It does so by interweaving two journeys made by the filmmaker: a journey by dog sledge in Greenland (north from Uummannaq in northwest Greenland) and a journey around the world (Tanzania, India and Costa Rica) in search of different places where poisonous (and frequently illegal) pesticides—such as the insecticide DDT, the agrochemical Nemagon and chemical waste—enter the local and global ecosystem. Aesthetically, this entanglement is signalled early in the film in a montage of industrial landscapes that is juxtaposed with a shot of two polar bears, while former Greenlandic minister Henriette Rasmussen explains the effects of global environmental pollution on the Arctic. This juxtaposition is repeated in a series of dissolves as the film moves back and forth between the filmmaker's two journeys. In the first of these, a panoramic shot of the Greenlandic coast, filmed from the frozen ocean, dissolves into a panoramic view of an open landscape in Tanzania (Fig. 2). The horizontal compositions and the mountains in the background of both shots create visual links between two very different environments. This anticipates the cultural links the filmmaker finds with

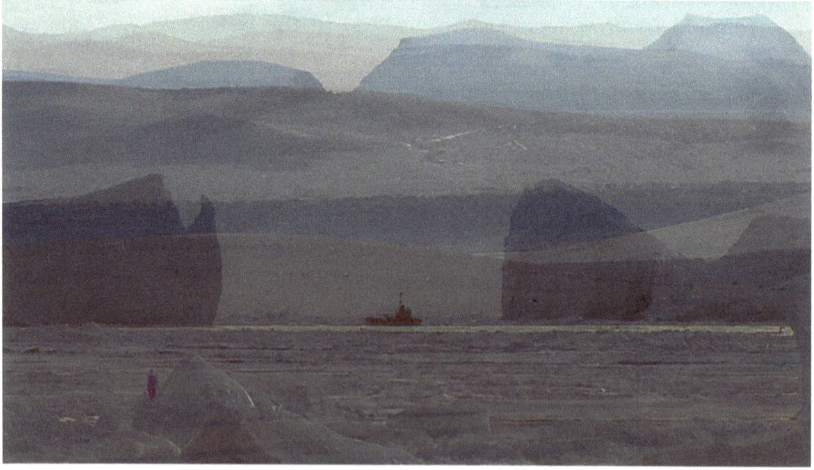

Fig. 2 Planetary entanglements from the Arctic to Tanzania: film still from Pipaluk Knudsen-Ostermann and Jan van den Berg, *Silent Snow* (2011)

the Indigenous Masai, among them a young woman who is, like Knudsen-Ostermann, an environmental activist and filmmaker. The Arctic and African landscapes are thus literally superimposed on each other, a filmic strategy that recurs as the film cross-cuts between the Arctic and other locations around the world. Indeed, the cross-cutting changes our perspective on both the former and the latter. The dog sledge journey over the Arctic ice also becomes a journey across a global archive of invisible pollution, made visible by the other journey.

The global flow of pesticides is visually signalled by the presence of water in different states of matter in several of the narrative transitions. At the beginning of the Indian segment, two dog teams moving on the Greenlandic ice dissolve into the Periyar River in Kerala, poisoned by chemical waste illegally piped into the river by countless factories. The near whiteout of the ice and the sky in the first shot is visually echoed by the hazy grey of the second. In the transition between the Arctic journey and the Costa Rican segment, a nighttime shot of an Inuk girl on the ice dissolves into another river landscape. In the last narrative transition, an Indigenous Bribri man showering in a Costa Rican waterfall dissolves into one of the dog teams moving on the Greenlandic ice. Only the Tanzanian segment has no water on the African side of the dissolve; an apt choice as we learn that the Masai are struggling with drought. In these shots, the use of water works on both literal and figurative levels; while it provides an apt image for the interconnectedness of the global ecosystem—reinforced by the 'fluid' cinematic technique of the dissolve—the stories we hear make it clear that the water we see also literally transports the pesticides. Overall, the aesthetic links the film creates between the Arctic and the other shooting locations contribute to a cultural ecology of planetary circulation—a circulation the filmmakers participate in through their own movements across continents.

In addition to signalling planetary entanglements, some of the dissolves explore parallels between human bodies and landscapes.[7] Thus, the aforementioned dissolve from the Inuk girl to the river in Costa Rica superimposes a medium close-up of her face on the landscape. Just before the dissolve, we hear Knudsen-Ostermann's voice explaining that "even in the glaciers pesticides are stored; as the ice melts, these will enter the

[7] I thank Anna Westerstahl Stenport, Scott MacKenzie and Klaus Dodds as well as the students of the lecture "Imagining the Arctic" (University of Zurich, spring 2017) for an inspiring discussion of *Silent Snow* that influenced the writing of the following paragraphs.

environment." The voice-over establishes the apparently pristine—but, by implication, polluted—river as a source of pesticides in Arctic snow and ice, and the superimposition of the human face onto the landscape portrays that the human body is affected by the same pollution. In the transition from the Indian segment back to Greenland, a close-up of Knudsen-Ostermann's face dissolves into a panorama of snow-covered ice at dusk, again presenting the body as a landscape and vice-versa, right after a discussion between the filmmaker and an Indian environmental activist about the pregnancy risks caused by pesticides. Indeed, the health hazards particularly to pregnant women and their children in the Arctic through consuming fish and (seal or whale) blubber are repeatedly foregrounded in the film. Bodies and landscapes are both presented as porous and interconnected with the ecosystem, with pesticides entering both. The titular snow thereby figures this intrusion into the body of the land, as the filmmaker's voice-over states at the beginning and very end of the film: "But now, an invisible enemy is threatening our homeland. It silently comes down with the snow."

Connecting with the common conception of Arctic ice as an environmental archive, the film nonetheless markedly differs from Western climate science discourse. Firstly, while the latter typically foregrounds Arctic ice as a climate archive storing information about global temperature changes, the film presents snow and ice as archives of environmental pollution. Secondly, it places this Arctic archive in relation to another archive, that of the human body, notably the female body—though not exclusively, as the film also foregrounds the effects of pesticides on male bodies, for example Costa Rican banana plantation workers' sterility and impotence resulting from exposure to Nemagon.

These reproductive anxieties connected to human bodies are representative of the film's overall concern with the future and the role of collective social action in shaping it. While *Chasing Ice* focuses on an individual 'white saviour' figure, the polar-hero-turned-photographer, *Silent Snow* emphasises community and human networks. While the former relies on masculinist individualism and logocentric rationalism, the latter centres femininity and community. If the 'hard' masculinity of *Chasing Ice* is figuratively connected to ice, the 'silent snow' of van den Berg and Knudsen-Ostermann's film is connected to a 'soft' poetics grounded in femininity and communal relations, visually expressed in the dissolve from the filmmaker's face to the snow-covered ice. Indeed, the film starts with a series of female figures speaking for their communities and establishing connections with each other, although the later segments increasingly show men in similar roles as well. Without asserting any essentialised soft or fluid femininity,[8] the film's calm, contemplative shots of snow-covered surfaces and various waterways emphasise connection and social relations embodied in the film's female figures but ultimately not restricted to any gender.

The emphasis on snow and ice as facilitators of social relations thereby differs maximally from the rhetoric of conquest and spectacle in *Chasing Ice*. The individuals we see in the film are locally grounded and engaged in small fights, actively proposing and promoting concrete solutions to environmental problems, such as Indigenous practices of malaria prevention as alternatives to the use of insecticides. The end credits appropriately begin with the words "[b]ased on stories from", superimposing the names of Knudsen-Ostermann's key interlocutors on the Arctic landscape. Referencing the importance of storytelling in Indigenous epistemologies, the film presents knowledge as emerging from many distinct, but related, stories. The individuals presented in the film are connected to various formal and informal networks at different scales, including the Women of Bhopal and a Meso-American meeting of Indigenous communities. The transnational production process emphasises collectivity and networks. While the film was produced mostly with Dutch funding, it was made in collaboration with a multiplicity of global and local institutions and organisations, and it had country producers for each of the narrative segments. The film ends with an appeal "for taking action in your neighborhood" as the last shot, a mobile aerial view of the frozen ocean, fades to black, inviting the viewer to enter its network of active subjects.

Conclusion: Rethinking the Arctic Sublime in the Anthropocene

The final scene of *Silent Snow* returns us to the question of the Arctic sublime and its relevance in communicating Arctic environmental topics in aesthetic form. As we see the filmmaker and her group travelling on the ice, interspersed with shots of two polar bears, we hear her tell an Inuit story in voice-over:

[8] For a good overview of debates about femininity and fluidity, see Stephens (2014).

> Long ago, in ancient days, there was a mother polar bear and her kids. They had taken off their furs because they were not white anymore. They sat down rubbing and cleaning them and while doing so they praised their country. They were singing about its beauty and how they enjoyed the sun and the sea. Then they heard a pack of dogs barking, they ran behind a small hill, but as soon as they had reached the hill, the hunters came and killed the polar bears with their arrows and spears. The bears had betrayed themselves by their wonderful singing, and sometimes you can still hear their song, a remembrance of how fragile beauty can be.

At the end of the story, we see a mobile shot of the polar bears in the distance, moving slowly in a wide expanse of ice as we hear an oratorio by Carl Philipp Emanuel Bach. The view soon dissolves into an aerial shot with the camera slowly moving above the frozen ocean as the credits continue to appear. The increasingly panoramic shots aestheticise the vast expanse of ice in ways that could be called sublime, but the story—along with the initial close-ups of human faces and close views of polar bears—presents the ice as a space carrying both living beings and narratives. The cross-cutting between humans and polar bears draws attention to the entwinement of life forms in the Arctic, while the story presents the ice as a site of tension and conflict. More importantly for this chapter, the filmmakers' use of the story advocates a reflective and critical engagement with aesthetic concepts, not only pointing to the fragility of beauty in times of environmental crisis but also suggesting that aesthetics can come at a price. *Silent Snow* thereby does not dismiss the value of aesthetic experience, but rather implies that aesthetic contemplation does not exist in a separate sphere of pure art—it is implicated in political and ideological struggles. The film encourages us to ask what hides behind aesthetic surfaces, including the environmental pollution caused by industrial capitalism, thinly veiled behind contemplative images of Arctic ice.

As such, the use of the sublime in *Silent Snow* directly contrasts with that of *Chasing Ice*. Where Orlowski's film paints a sublime spectacle of Nature with a capital N, located in a realm disconnected from humans, van den Berg and Knudsen-Ostermann's Arctic sublime emphasises precariousness by taking human experiences as the starting point for tracing ecological entanglements. The environmentalism of *Silent Snow* emerges from human practices in space; it presents many little stories rather than one universal story focalised through "one man's mission to change the tides of history", to cite the tagline of *Chasing Ice*. It is significant that

Silent Snow does not present itself as a story about climate change—indeed, the term is never mentioned in the film. While the effects of global warming are repeatedly alluded to, they are mentioned in relation to other environmental problems caused by industrial capitalism, such as the link between melting glaciers and the global circulation of pesticides. *Chasing Ice* plays into the universalist rhetoric of the Anthropocene that has been critiqued by postcolonial theorists and Indigenous scholars (see, for instance, Buckland 2018; Todd 2015). Its logocentric, masculinist communicative impulse aligns with what scholars like Gayatri Chakravorty Spivak and Dipesh Chakrabarty refer to as globalism, or the top-down vision of a homogeneous world connected to a universal subject (Chakrabarty 2021; Spivak 2003, 71–102). The bottom-up, grassroots activism of *Silent Snow*, on the other hand, is rooted in an inclusive femininity that emphasises situated knowledges and collective action through planetary relations without claims to totality. Its Indigenous transnationalism and cosmopolitanism (see, among others, Forte 2002) contrast with the former film's hegemonic globalism, encapsulated in the contrast between soft snow and hard ice.[9]

The reference to polar bears is thereby representative of *Silent Snow*'s emphasis on active and resilient subject positions within an interconnected ecosystem. Very different from the iconic use of polar bears as emblems of a 'dying Arctic' in Western environmental communication, the film draws on Inuit conceptions of the polar bear "as a symbol of the resilience, patience and determination that is needed to survive in the harsh climate" (*Inuvialuit and Nanuq* 2015). While the Arctic sublime of *Chasing Ice* posits—against its stated intentions—a passive spectator of an environmental disaster conceived in apocalyptic terms, *Silent Snow* answers the circulation of pesticides with a human planetary network within a relational environment. Both films communicate environmental knowledge through aesthetic form and mobilise a version of the sublime to do so. While the sublime of both films transcends the space of the nation and is connected to global environmental processes that remain largely intangible, *Silent Snow* works with what we might think of as an Indigenous

[9] As Kirsten Thisted points out, "Greenlanders are actively engaged in forming a cosmopolitan and urban modernity that reflects the range of globalisation processes" (2014, 98). Somewhat surprisingly, however, she refers to *Silent Snow* as an example of a film that presents Greenland as remote and 'traditional', a claim that is somewhat hard to reconcile with the film's presentation of Knudsen-Ostermann as a cosmopolitan travelling the world and fluently speaking four languages (Kalaallisut, Danish, English and Spanish).

sublime grounded in Inuit environmental philosophy, although it mixes cultural registers. Like Lynge's story about icebergs, the final scene of the film combines Western and Indigenous references as the pathos of Bach's music coexists with an aesthetic exploration of the Inuit environmental concept of *sila* through aerial shots.[10] Whether the sublime can indeed be meaningfully connected to Indigenous environmental aesthetics is better answered by Indigenous scholars, but as this chapter has shown, the concept remains relevant in contemporary Arctic environmental imaginaries, for better or worse.

Bibliography

Barsam, Richard. 1992. *Nonfiction Film: A Critical History*. Rev. and exp. ed. Bloomington: Indiana University Press.

Baudry, Jean-Louis. 1974–1975. Ideological Effects of the Basic Cinematographic Apparatus. Trans. Alan Williams. *Film Quarterly* 28 (2): 39–47. https://doi.org/10.2307/1211632.

Belanger, Noelle. 2019. "It Looks Like the Surface of the Moon": The Arctic Landscape in the Western Imaginary. In *Arctic Cinemas and the Documentary Ethos*, ed. Lilya Kaganovsky, Scott MacKenzie, and Anna Westerstahl Stenport, 215–230. Bloomington: Indiana University Press. https://doi.org/10.2307/j.ctvcj2wqq.5.

Bloom, Lisa. 1993. *Gender on Ice: American Ideologies of Polar Expeditions*. Minneapolis: University of Minnesota Press.

———. 2022. *Climate Change and the New Polar Aesthetics: Artists Reimagine the Arctic and Antarctic*. Durham: Duke University Press.

Buckland, Adelene. 2018. "Inhabitants of the Same World": The Colonial History of Geological Time. *Philological Quarterly* 97 (2): 219–240.

Burke, Edmund. 1990 [1757]. *A Philosophical Enquiry*. Oxford: Oxford University Press.

Chakrabarty, Dipesh. 2021. *The Climate of History in a Planetary Age*. Chicago: The University of Chicago Press.

Cone, Marla. 2005. *Silent Snow: The Slow Poisoning of the Arctic*. New York: Grove Press.

Dodds, Klaus. 2018. *Ice*. London: Reaktion Books.

Dodds, Klaus, and Rikke Bjerg Jensen. 2019. Documenting Greenland: Popular Geopolitics on Film. In *Arctic Cinemas and the Documentary Ethos*, ed. Lilya

[10] The use of mobile aerial shots as a cinematic rendition of *sila* is common in Inuit cinema. Other examples include the drone shots in Marc Fussing Rosbach's *Akornatsinniittut—Tarratta Nunaanni* (2017) and Lucy Tulugarjuk's *Tia and Piujuq* (2018).

Kaganovsky, Scott MacKenzie, and Anna Westerstahl Stenport, 155–174. Bloomington: Indiana University Press. https://doi.org/10.2307/j.ctvcj2wqq.5.

Dodds, Klaus, and Mark Nuttall. 2016. *The Scramble for the Poles*. Cambridge: Polity.

Farish, Matthew. 2006. Frontier Engineering: From the Globe to the Body in the Cold War Arctic. *The Canadian Geographer / Le Géographe canadien* 50: 177–196. https://doi.org/10.1111/j.0008-3658.2006.00134.x.

Flaherty, Robert. 1922. *Nanook of the North*, 79 minutes.

Forte, Maximilian C., ed. 2002. *Indigenous Cosmopolitans: Transnational and Transcultural Indigeneity in the Twenty-First Century*. New York: Peter Lang.

Hansson, Heidi, and Cathrine Norberg. 2009. Revisioning the Value of Cold. In *Cold Matters Cultural Perceptions of Snow, Ice and Cold*, 7–22. Umeå: Umeå University and the Royal Skyttean Society.

Hastrup, Kirsten. 2013. The Ice as Argument: Topographical Mementos in the High Arctic. *Cambridge Anthropology* 31 (1): 51–67. https://doi.org/10.3167/ca.2013.310105.

Hilderbrand, Lucas. 2020. On Nature Programming, the Anthropocene, and the Humanitarian Impulse in Documentary. *Cambridge Journal of Postcolonial Literary Inquiry* 7 (2): 210–215. https://doi.org/10.1017/pli.2019.35.

Hill, Jen. 2008. *White Horizon: The Arctic in the Nineteenth-Century British Imagination*. Albany: State University of New York Press.

Inuvialuit and Nanuq: A Polar Bear Traditional Knowledge Study. 2015. Inuvik: Joint Secretariat, Inuvialuit Settlement region.

Kaganovsky, Lilya, Scott MacKenzie, and Anna Westerstahl Stenport, eds. 2019a. *Arctic Cinemas and the Documentary Ethos*. Bloomington: Indiana University Press. https://doi.org/10.2307/j.ctvcj2wqq.5.

———. 2019b. Introduction: The Documentary Ethos and the Arctic. In *Arctic Cinemas and the Documentary Ethos*, 1–28. Bloomington: Indiana University Press. https://doi.org/10.2307/j.ctvcj2wqq.5.

Körber, Lill-Ann. 2017. Toxic Blubber and Seal Skin Bikinis, or: How Green is Greenland? Ecology in Contemporary Film and Art. In *Arctic Environmental Modernities*, ed. Lill-Ann Körber, Scott MacKenzie, and Anna Westerståhl Stenport, 145–167. Cham: Palgrave.

Loomis, Chauncey C. 1977. The Arctic Sublime. In *Nature and the Victorian Imagination*, ed. U.C. Knoepflmacher and G.B. Tennyson, 95–112. Berkeley: University of California Press. https://doi.org/10.1525/9780520340152-008.

Lynge, Aqqaluk. Strengthening Culture through Change: Will Climate Change Strengthen or Destroy Us? Address to the Royal Scottish Geographical Society at the University of Edinburgh, Inuit Circumpolar Council, 19 October 2009. http://inuit.org/our-work/climate-change/aqqaluk-lynges-speech-at-edinburgh-university/?contUid=0. Accessed 26 Aug 2020.

Martin, Keavy. 2012. *Stories in a New Skin: Approaches to Inuit Literature.* Winnipeg: University of Manitoba Press.
Massey, Doreen. 2005. *For Space.* London: SAGE.
Morgan, Benjamin. 2016. After the Arctic Sublime. *New Literary History* 47 (1): 1–26.
Morton, Timothy. 2007. *Ecology Without Nature: Rethinking Environmental Aesthetics.* Cambridge, MA: Harvard University Press.
Orlowski, Jeff. 2011. *Chasing Ice*, 74 mins.
Potter, Russell A. 2007. *Arctic Spectacles: The Frozen North in Visual Culture, 1818–1875.* Seattle: University of Washington Press.
Qitsualik, Rachel A. 2013. Inummarik: Self-Sovereignty in Classic Inuit Thought. In *Nilliajut: Inuit Perspectives on Security, Patriotism and Sovereignty*, ed. Scot Nickels, Karen Kelley, Carrie Grable, Martin Lougheed, and James Kuptana, 23–34. Ottawa: Inuit Tapiriit.
Renov, Michael. 2019. Documenting the Arctic Sublime. In *Arctic Cinemas and the Documentary Ethos*, ed. Lilya Kaganovsky, Scott MacKenzie, and Anna Westerstahl Stenport, 207–214. Bloomington: Indiana University Press. https://doi.org/10.2307/j.ctvcj2wqq.
Sanguya, Joelie, and Shari Gearhead. 2010. Preface. In *SIKU: Knowing Our Ice*, ed. Igor Krupnik, Claudio Aporta, Shari Gearheard, Gita J. Laidler, and Lene Kielsen Holm, ix–x. Dordrecht: Springer.
Spivak, Gayatri Chakravorty. 2003. *Death of a Discipline.* New York: Columbia University Press.
Steinberg, Philip, and Berit Kristoffersen. 2017. "The ice edge is lost . . . nature moved it": Mapping Ice as State Practice in the Canadian and Norwegian North. *Transactions of the Institute of British Geographers* 42: 625–641. https://doi.org/10.1111/tran.12184.
Stephens, Elizabeth. 2014. Feminism and New Materialism: The Matter of Fluidity. *Inter/alia: A Journal of Queer Studies* 9: 186–202. https://doi.org/10.51897/interalia/BOSZ9092.
Thisted, Kirsten. 2014. Cosmopolitan Inuit: New Perspectives on Greenlandic Film. In *Films on Ice: Cinemas of the Arctic*, ed. Scott MacKenzie and Anna Westerståhl Stenport, 97–104. Edinburgh: Edinburgh University Press. https://doi.org/10.3366/edinburgh/9780748694174.003.0007.
Todd, Zoe. 2015. Indigenizing the Anthropocene. In *Art in the Anthropocene Encounters Among Aesthetics, Politics, Environments and Epistemologies*, ed. Heather Davis and Etienne Turpin, 241–254. London: Open Humanities Press. https://doi.org/10.26530/OAPEN_560010.
Truscello, Michael. 2018. Catastrophism and Its Critics: On the New Genre of Environmentalist Documentary Film. In *Interrogating the Anthropocene: Ecology, Aesthetics, Pedagogy, and the Future in Question*, ed. Jan Jagodzinski, 257–275. Cham: Palgrave.

van den Berg, Jan, and Pipaluk Knudsen-Ostermann. 2011. *Silent Snow*, 71 mins.

Watt-Cloutier, Sheila. 2002. Wake-up Call. *Biodiversity* 22 (3): 39–40.

Westerstahl Stenport, Anna, and Richard S. Vachula. 2017. Polar Bears and Ice: Cultural Connotations of Arctic Environments that Contradict the Science of Climate Change. *Media, Culture & Society* 3 (3): 282–295. https://doi.org/10.1177/0163443716655985.

Frozen-Ground Cartoons—Revealing the Invisible Ice

Frédéric Bouchard and Ylva Sjöberg

> The science community tends to treat itself very seriously and quickly becomes prickly if criticised from outside. This is perhaps not a very good thing, and therefore, science needs cartoons. (Dominiczak 2017, 934)

Prologue

It is hot. People are probably sweating outside. It is summertime of 1844 in the city of Yakutsk in Siberia, but it surely does not feel like it in there. In there? In the darkness, nearly one hundred metres deep below the surface, uncomfortably cramped in a wooden basket balancing between the walls of a narrow well, a man tries to read, by candlelight, the thermometer which he just retrieved from a hole drilled in one of the walls.

F. Bouchard (✉)
University of Sherbrooke, Sherbrooke, QC, Canada

Y. Sjöberg
University of Copenhagen, Copenhagen, Denmark

A. Hemkendreis, A.-S. Jürgens (eds.), *Communicating Ice through Popular Art and Aesthetics*, Palgrave Studies in Media and Environmental Communication,
https://doi.org/10.1007/978-3-031-39787-5_12

Minus five degrees Celsius. Clearly, there is no running water in there. It is totally frozen.

Alexander von Middendorff, explorer and naturalist, soon-to-be member of the Imperial Saint Petersburg Academy of Sciences (now the Russian Academy of Sciences), is much more captivated than discouraged by such weird news. By boat, sled, horse, reindeer or foot, he has been exploring the cold landscapes of Siberia for several years now, having covered about 20,000 kilometres (Tammiksaar and Stone 2007). He barely survived one particularly harsh late-fall trip across the Taymyr Peninsula—the northernmost part of the Eurasian mainland, lying between two majestic 'monsters' of the Arctic: the Yenissei and the Lena rivers. Still, he is obsessed with this new phenomenon reported from northern and eastern Siberia: the 'never-melting ground ice'. Ice that never melts? Is that possible? Maybe. This is why he and his team spent the year in the so-called Shergin Shaft—a well that local merchant Fedor Shergin dug over a ten-year period (1827–1837) to try to find drinkable water in the Lena area (MPI 2015). Unsuccessfully. But the absence of liquid water spurred scientific interest in ground ice, including by the Tsar himself, Nicholas I. Soon enough, von Middendorff would be known as a pioneer in the science of вечная мерзлота (*vechnaya merzlota*), the eternally frozen soil.

Nearly two centuries later, in the early twenty-first century, frozen ground would become a dynamic field of scientific research, with major implications for local Arctic communities and global climate. We call this interdisciplinary field of knowledge Geocryology, the science of ground ice (Shiklomanov 2005). Communicating the importance of Geocryology is challenging because it is hidden within the earth: How do you visualise something invisible?

The World of Underground Ice

If you fly over central Yakutia (eastern Siberia), you will see a landscape that is both impressive and calming (Fig. 1). Not a single mountain, no ocean, no sea, no large city—except Yakutsk, with its 300,000 people, but everywhere on the horizon: the boreal forest (taïga), lakes—lots of lakes, tiny and big, shallow and deep, dark brown, milky beige or blue-green, maybe some roads, electricity lines and villages here and there? But other than that: an amazingly flat landscape, the horizon, the sky.

But there is a lot more in there. You will have to get out of the aircraft and walk a little—or sometimes a lot—to see it. Something practically

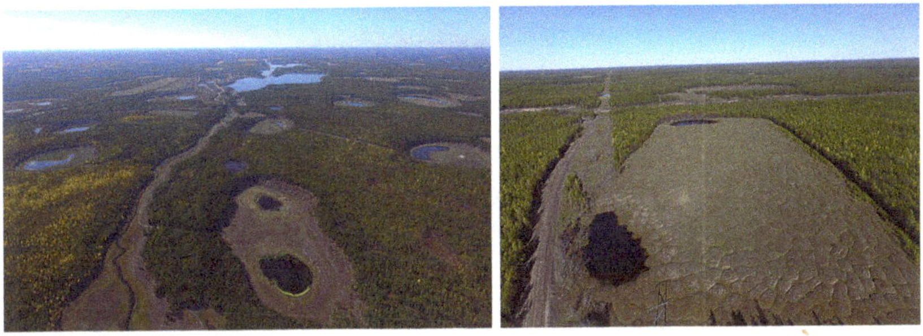

Fig. 1 Typical photos of Central Yakutia viewed from the air, 2018 © A. Séjourné

invisible to the non-exercised eye, lying just beneath your feet, has been forming, kind of breathing, sleeping there for tens of thousands of years. It underlies the land, several hundreds of metres thick, millions of square kilometres across. It is present everywhere, except for underneath the majestic Lena River, but of course you cannot see it. It is nevertheless the foundation of northern ecosystems and infrastructure. It is the hidden world of underground ice. It is frozen ground. It is permafrost.

Occupying more than 20 million square kilometres of the Earth's high-latitude and high-altitude landscapes, permafrost is not only a key component of our global climate, but it is also arguably the most overlooked, invisible part of the cryosphere (from the Greek '*krios*', cold or frost), or the world of ice in its many forms (e.g. snow, glaciers, ice caps, sea ice, lake and river ice). On a map of the northern hemisphere centred on the North Pole (Brown et al. 2002), it spreads across half of Canada, most of Alaska and Siberia, the Tibet plateau and some fringes of northern Scandinavia and Greenland (Fig. 2). The closer you are to the pole, the more widespread or continuous, colder and thicker it is—up to a kilometre thick in the Verkhoyansk Range, in eastern Siberia (!). More south, it is discontinuous and pretty thin—only a few metres in northern Scandinavia, for example. Its geographic distribution more or less follows Arctic and sub-Arctic climatic zones, from the polar desert to the tundra and the boreal forest.

There is a reason for this: permafrost definition is strictly a thermal one. It is defined as any soil or rock with a temperature at or below 0 °C for at least two consecutive years (Van Everdingen 2005). Said otherwise, it is the temperature at which fresh water normally freezes/melts. Nothing

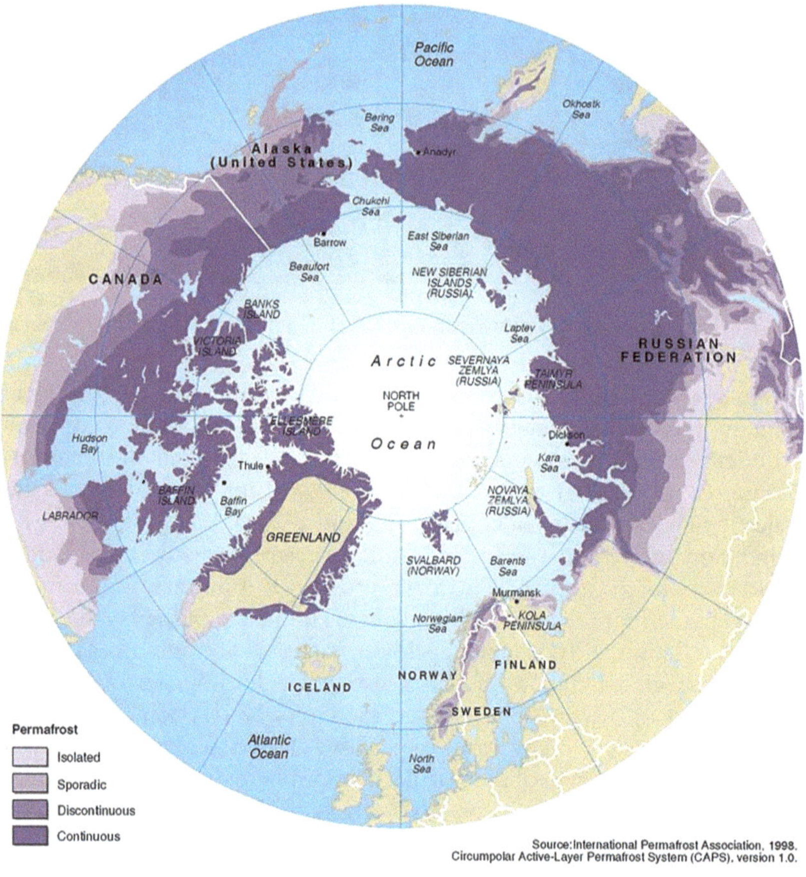

Fig. 2 Circum-Arctic map of permafrost, in Circum-Arctic map of permafrost and ground-ice conditions, Brown et al. (2002) © International Permafrost Association 1998

more, nothing less. It means that permafrost can be just any cold rock, granite or limestone, but can also be frozen clay or peat soil with lots of ice lenses—vertical and horizontal, each a few millimetres to several metres thick. It also suggests that permafrost can contain air, gas bubbles, bacteria and even liquid water—salty water will not freeze above about -2 °C. In other words, it can be ice-rich, but it is not just ice. Meaning: when ground temperature rises above 0 °C, permafrost does not melt, as pure ice does. Instead, like a frozen rye bread—it thaws.

Nearly five million people live in permafrost areas across the Arctic (Ramage et al. 2021). Just like ice sheets in Greenland and Antarctica, or mountain glaciers around the world—often depicted in the general media—permafrost regions are highly sensitive to climate change and human activities (Fig. 3). In fact, these regions are already affected by widespread thawing (Biskaborn et al. 2019). Locally, permafrost thaw can jeopardise the stability of infrastructure—buildings, roads, air strips, railways and so on. Such severe damage might affect as much as one third of total pan-Arctic infrastructure by mid-century (Hjort et al. 2018). It can also threaten natural ecosystems, some of which play a key role in Indigenous traditional way of life—hunting, fishing and trapping grounds, berry harvesting and so on. But permafrost thaw also has a global connection. Formerly trapped/frozen organic matter can be released in the environment and eaten by certain microbes producing greenhouse gases (CO_2, CH_4), thereby amplifying ongoing warming (Schuur et al. 2015). Permafrost research is therefore of considerable scientific and societal importance.

How can we make sure that stakeholders and the public are aware of that? How can we bring communities together and enhance resilience to the multifaceted consequences of climate warming across permafrost regions?

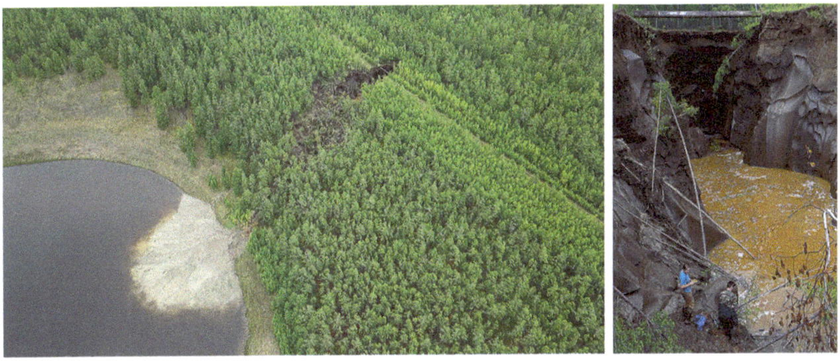

Fig. 3 Field example of ice-rich permafrost thaw in Central Yakutia, generating rapid transfers of meltwater, organic matter and sediments, 2019 © A. Séjourné (**a**) and L. Jardillier (**b**)

Well, this is not a small task, especially when working with something invisible. It thus requires the combined creative skills of the scientists who produce 'permafrost facts', and the artists who communicate these facts.

Permafrost Science Communication: *Frozen-Ground Cartoons*

The *Frozen-Ground Cartoons* (FGC) project was born in 2015–2016 with the urge to make permafrost science accessible and fun for the general public, especially for school kids, students and teachers. Our vision was, and still is, to connect arts and science along visual, narrative and metaphoric axes (Farinella 2018). We also want to communicate the relevance of the discipline in a global context in an engaging manner (Bouchard et al. 2019, 2022).

FGC started with a group of motivated early-career researchers (ECR) who received 5000 Euros from the International Permafrost Association (IPA) to make comic strips about permafrost. We knew a lot about permafrost but much less about art and therefore decided to spend all the money on artists—a great decision, in retrospect. Step one was a call for artists, which resulted in an overwhelming response. We received applications from 49 artists in 15 countries, working within a wide range of styles and formats. Some were humorous, some were breathtakingly beautiful, some were interactive. Ten artists were invited to prepare and submit a one-page pitch (Fig. 4). Four of the pitches featured ducks, three featured reindeer and one featured a polar bear. Most importantly, all featured permafrost centre stage. Permafrost from above, from space, as something hidden, as something you can touch, hold in your hands or stand on as an invisible stage. Permafrost that affects landforms, vegetation, wildlife habitat and infrastructure. Permafrost as a patient with a fever, or as melting ice cream. Even the sound of permafrost when you hit it with a steel rod—'klonk' rather than 'klink'—was showcased. And they featured the researchers measuring, observing, poking and sampling permafrost. Our greatest regret at this stage was that we only had enough money to continue working with two of the artists.

The one-page pitch *Walking on a Giant*, by the Finnish artist Heta Nääs, creatively depicted the great depths of permafrost with scientific detail and accuracy (Fig. 4a). Canadian artist Noémie Ross used humour and metaphors to explain the urgency of permafrost research (Fig. 4b).

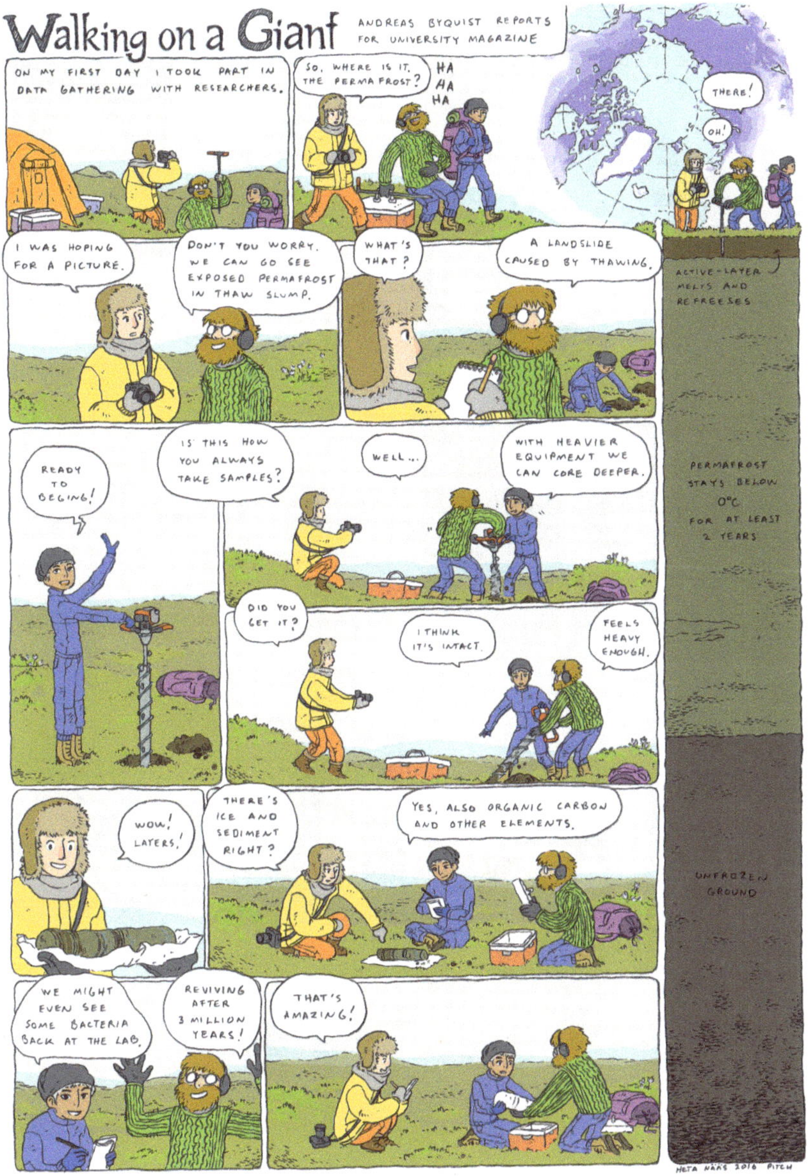

Fig. 4 'Samples' of the one-page pitches by Heta Nääs (**a**) and Noémie Ross (**b**), which were selected for the project, 2017 © Heta Nääs and Noémie Ross

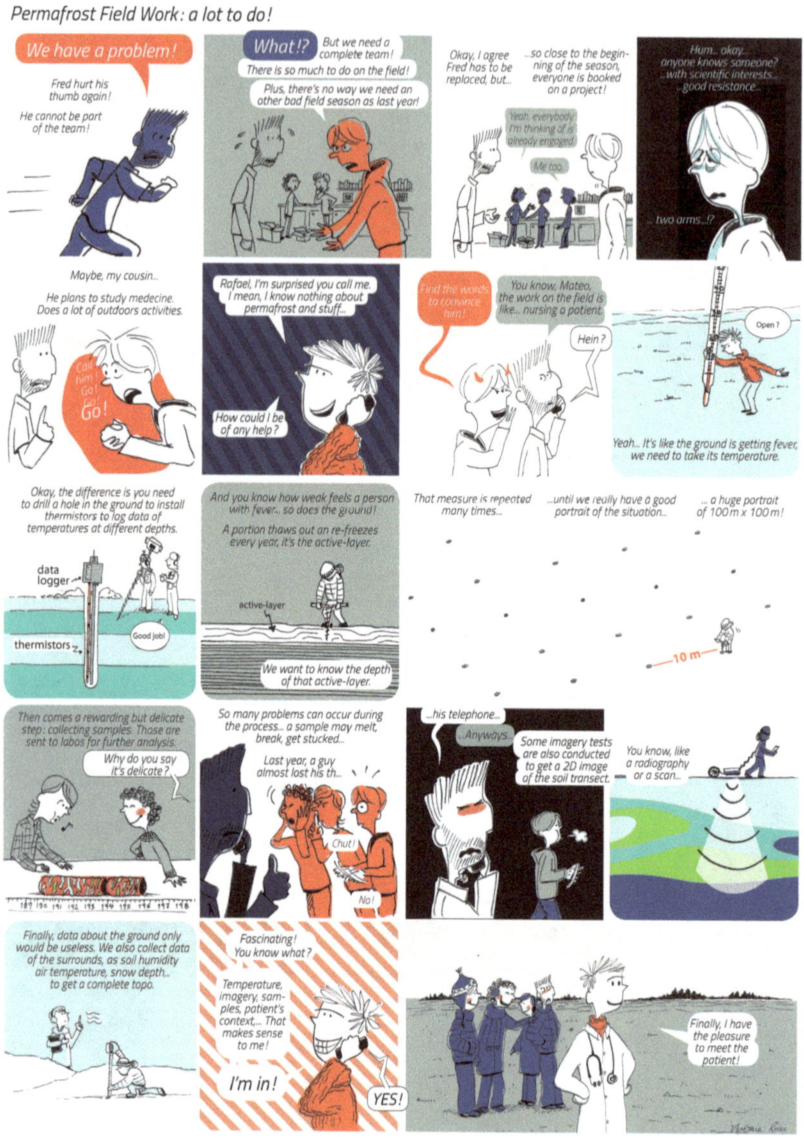

Fig. 4 (continued)

These two artists, selected to produce the final FGC, used very different approaches to engage readers and communicate our science. Heta and Noémie produced beautiful and impactful permafrost comics in contrasting styles. As natural scientists, we were immediately drawn to Heta's artistic ability to depict the scale and perspective of underground features. Through discussions with our team of ECR, Heta was able to create extremely accurate drawings of a vast range of permafrost features, processes and scientific methods. She communicated the invisible by using a car for scale when visualising the great thickness of permafrost and ice wedges or the scale of a thaw slump. And all the drawings of scientists and animals in the field included clouds of mosquitoes. Noémie's approach was completely different. She produced five short stories leveraging humour, metaphors, emotion and very cute animal characters. In her pitch, a medical student joins a team of permafrost researchers for field work on the tundra and diagnoses the thawing ground as a patient with a fever. In another story, a reindeer tries to convince her calf to wear arm floaties to avoid drowning in mud resulting from thawed permafrost.

Our ECR group provided the artists with a starting pack, with basic information about permafrost, field-based research and the urgency of permafrost research at local and global scales. To help the artists develop the full FGC comics, we provided input and feedback on scientific aspects of their work. We answered questions such as "Could a thaw slump look like this?", "Could it be located there?", "Would researchers be sampling mud there and then?", "What equipment would be used for that?", "Could I add a polar bear here?" ("—Yes!"). With team members spread over two continents, most of this communication took place online and included heated discussions about how the artists could best fit all our science into a limited number of pages of art.

The FGC 28-page comic book was released in English in 2017 and includes the two ten-page comics and the one-page pitches from Heta and Noémie, as well as an illustrated permafrost glossary and a circumpolar map of permafrost in the Arctic (Nääs et al. 2017). Versions in Swedish, German, French, Russian, Inuktituk, Danish, Greenlandic (Kalaallisut) and even Luxembourgish are now also available (Fig. 5). The Swedish version includes one extra page of illustrations about permafrost in Scandinavia, while the Greenlandic and Danish versions include four pages of extra material about permafrost research in Greenland. All comics were made available on a website (Frozen-Ground Cartoons 2017a), with a Creative Commons Attribution-Non Commercial-ShareAlike

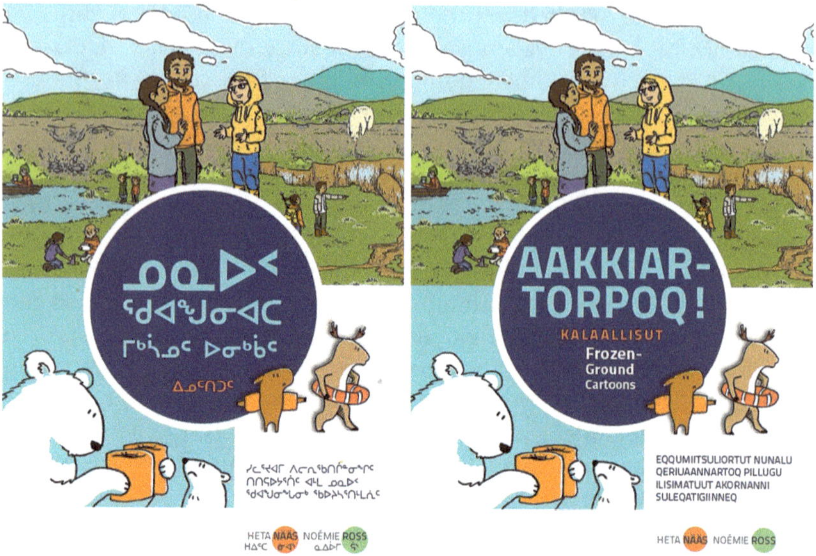

Fig. 5 Examples of cover pages of FGC translations, including in Inuktitut (**a**) and Greenlandic/Kalaallisut (**b**), 2020 © Heta Nääs and Noémie Ross

international licence to ensure that material will be available for anyone who wants to share information about permafrost research.

The FGC team is extremely proud of the FGC comics. But the comics were never made for us—at least, not us alone. How would our audience react to this new material? We imagined kids sitting in classrooms or at home reading our comics and being blown away by the *cool* science stories. But how would we know, if we were not there when they read it? Beyond the statistics of downloads from our website, it is a challenging task to gauge the impact of the cartoons. However, some live outreach events have provided opportunities for valuable feedback. At the Bolin Centre Climate Festival for Schools in Stockholm for example, we handed out printed copies of the Swedish version to school kids (ages 10–16) and we added a quiz about permafrost and about the comics (for those who had read it). The students indicated that the comics were exciting, educational and fun; the students who had read the comics also scored higher on the quiz. In an alternative approach to engage the students with our comics,

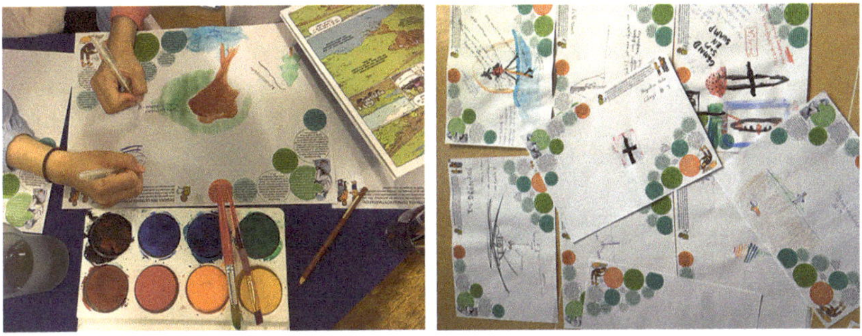

Fig. 6 Ylva Sjöberg, school drawing context ('design the best permafrost sampling tool'), 2017

we supplied watercolours and papers and challenged the students to design the best possible instrument for a permafrost researcher to bring to the field (Fig. 6). Full of ideas, the students suggested, for instance, genetically modified trees and mushrooms for logging temperature, drone-based instruments and even a rifle that shoots sensors into the permafrost. Many based their drawings and depictions of permafrost on art from the cartoons and included underground features as well as specific landforms, such as thaw slumps. The winning contribution showed a temperature logger with sensors spreading like a root network to capture temperature at different depths and locations in the ground.

Fast-forward to autumn 2019. We presented 'permafrost workshops' to different high schools across Luxembourg (ages 14–18) for a week. Featuring the FGC and related 'by-products' (see below), these activities proved that there is strong interest among the younger generations for the world of underground ice. Permafrost-mania in Luxembourg? Kind of. These frozen, amazing, almost greater-than-nature landscapes located at the other end of the world seem to exert a mysterious attracting force, something maybe not so different from the sublime impression left by awe-inspiring icy features in polar regions—glaciers, ice caps, sea ice—but in that case: hidden and only perceivable when visualised! As scientists, we are often amazed by the vastness and beauty of Arctic landscapes, and in their unique, artistic way, FGC can help us share such feelings and at the same time make our science more 'real' to students and the general public.

The *Frozen-Ground Cartoons* soon moved well beyond comics (Bouchard et al. 2019). Augmented reality material has been added to the comics via an app so that the reader can project—directly from the comics with a phone and/or tablet—animated objects like a three-dimensional map, a reindeer, a chainsaw and so on, as well as five short information videos. For those who still have not had enough permafrost in their lives, we developed a board game. Up to six players can spend an evening answering permafrost science questions in a trivial-pursuit style game—including six categories of questions covering both natural and human sciences. Soon we hope that we can introduce our audience to the sounds of permafrost research (an audiobook? a podcast?). There is even a developing project of immersive virtual reality (VR) experiences of the permafrost landscape. Imagine this VR synopsis: a group of kids from sun-baked California gets transported in Siberia and meets with another group of kids from the coldest city on Earth: Yakutsk, the cradle of permafrost science. We believe in the power of collaborative relationships between scientists and artists, and we love testing new ways of engaging the public in our (somewhat hidden) ice science. For many people, permafrost is something mysterious in a faraway Arctic which they have never visited. With the help of art, we attempt to in some way bring them there—to the land of permafrost—and let them experience the landscapes that fascinate us scientists.

An Ice Scientist's Perspective on Arts and Aesthetics

In addition to providing a multi-language, interdisciplinary outreach, communication and education product to the general public, the FGC project made 'us'—our small international community of permafrost geeks—work and collaborate in a new, different way. There have been some challenges, starting with the simple fact that, as mentioned several times in this chapter—including in the title—permafrost is the invisible ice. Both literally, hidden beneath our feet, and figuratively, located so far away in these inaccessible frozen lands. Yet, we wanted to transmit our passion for this frozen world by visualising the impact of permafrost on people and the natural environment in space and time. We simply had to team up with artists to tackle this challenge.

Drafting our 'application call', we had no idea about what kind of contributions or interactions we would have. Our main selection criteria were quite general: overall quality, originality, diversity, complementarity. But as natural scientists we faced a new challenge—How do you *objectively* evaluate 'overall quality'? We realised that we all valued different aspects of the artists' work and after lengthy discussions ended up with a democratic selection process. Nevertheless, the humour in Noémie Ross's cartoons and Heta Nääs's play with scales and perspective are aesthetic aspects of their work that directly spoke to us, as efficient ways of explaining permafrost science and conveying the 'joy of learning' new stuff. Our input during the process of creating the final cartoons was mostly limited to the scientific aspects of the work, leaving the artistic decision making (if such a distinction can be made) to the artists. We figured it best that we stuck to our area of expertise and let the artists decide on aesthetic communication methods. While this may sound straightforward, our demands for greater scientific details naturally created constant aesthetic challenges for the artists.

We like to think that our comics have made the fascinating underground realm of invisible ice a little bit more tangible for the reader of our *Frozen-Ground Cartoons*; that we have begun to provide some explanation about why it matters and why we study it as the world around us warms. When von Middendorff descended into Shergin shaft less than 200 years ago, very little was known about permafrost but since then the scientific literature has grown exponentially. We believe the time has come for similar growth in permafrost awareness and concern by the general public. Comics and arts are one way of facilitating this.

Epilogue

Back into the permafrost science now, or, as it was defined earlier: Geocryology. The world's deepest vertical ice tunnel had been buried and forgotten for more than half a century. But thanks to the efforts of motivated local residents and staff from the Melnikov Permafrost Institute (part of the Siberian Branch of the Russian Academy of Sciences), the 'Shergin Shaft' has recently been repaired and secured, more than 175 years after von Middendorff's visit and measurements. Watched by a curious crowd, three persons—a camerawoman, a rescue chief and a TV producer—are slowly moved down into the tunnel in November 2009, metre by metre (Bochkarev 2009). Maybe it says something about our entertainment-prone epoch: what had started in the mid-nineteenth century as a formal scientific investigation was now showcased to the residents as an amusing curiosity. Could this be a typical twenty-first-century way of visualising the 'historical interest' of *vechnaya merzlota*? Nonetheless, our three modern explorers in 2009, down in the shaft, see exactly what von

Middendorff and his colleagues had seen in the middle of the nineteenth century, in the exact same well, originally dug to find—with no success—drinkable water: a world of frozen soil and glittering ice, tens of thousands of years old.

The invisible, never-melting ground ice.

Acknowledgements This whole project started with enthusiastic support—both financial and philosophical—from the International Permafrost Association (IPA). We are very thankful to all the artists who submitted cartoon proposals at the beginning of the project, especially the two artists that were finally chosen to work with us: Heta Nääs and Noémie Ross. We are also extremely grateful to the awesome group of permafrost researchers and science communicators who invested their time in the project. They appear on FGC's website (Frozen-Ground Cartoons 2017b). We also warmly thank T. Gibéryen, C. Haugk and J. Lenz for their great help in preparing and conducting the series of permafrost workshops in Luxembourg during the autumn of 2019. Finally, we wish to thank all the people and organisations who helped in translating the comics into several other languages (D. Nekrasov, K. Rasmussen, K. Langley and Ayaya Marketing and Communication).

Bibliography

Biskaborn, Boris K., Sharon L. Smith, and Jeannette Noetzli, et al. 2019. Permafrost Is Warming at a Global Scale. *Nature Communications* 10 (1): 264:1–11. https://doi.org/10.1038/s41467-018-08240-4.

Bochkarev, Bolot. 2009. Descending into Shergin's Shaft, the World's Deepest in the Permafrost Zone. *YouTube*. https://youtu.be/e256rh8OImA. Accessed 21 Nov 2022.

Bouchard, Frédéric, Julie Sansoulet, Michael Fritz, Julie Malenfant-Lepage, Alexandre Nieuwendam, Michel Paquette, Ashley C.A. Rudy, Matthias B. Siewert, Ylva Sjöberg, J. George Tanski, Otto Habeck, and Jon Harbor. 2019. "Frozen-Ground Cartoons": Permafrost Comics as an Innovative Tool for Polar Outreach, Education, and Engagement. *Polar Record* 54 (5–6): 366–372. https://doi.org/10.1017/S0032247418000633.

Bouchard, Frédéric, Michael Fritz, and Ylva Sjöberg. 2022. Redrawing Permafrost Outreach. *Nature Reviews Earth & Environment* 3: 7. https://doi.org/10.1038/s43017-021-00255-8.

Brown, Jerry, J. Oscar Ferrians, J. Heginbottom, and E. S. Melnikov. 2002. *Circum-Arctic Map of Permafrost and Ground-Ice Conditions, Version 2*. Boulder: National Snow & Ice Data Center. https://doi.org/10.7265/skbg-kf16. Accessed 26 July 2022.

Dominiczak, Marek H. 2017. Cartoons: Another Long-Standing Bridge Between Science and the Arts. *Clinical Chemistry* 63 (4): 934–935. https://doi.org/10.1373/clinchem.2016.266783.

Farinella, Matteo. 2018. The Potential of Comics in Science Communication. *Journal of Science Communication* 17 (1): Y01:1–17. https://doi.org/10.22323/2.17010401.

Frozen-Ground Cartoons. 2017a. Home. www.frozengroundcartoon.com. Accessed 21 Nov 2022.

———. 2017b. Team. www.frozengroundcartoon.com/team/. Accessed 21 Nov 2022.

Hjort, Jan, Olli Karjalainen, Juha Aalto, Sebastian Westermann, Vladimir E. Romanovsky, Frederick E. Nelson, Bernd Etzelmüller, and Miska Luoto. 2018. Degrading Permafrost Puts Arctic Infrastructure at Risk by Mid-Century. *Nature Communications* 9 (1): 5147:1–9. https://doi.org/10.1038/s41467-018-07557-4.

Melnikov Permafrost Institute (MPI). 2015. Shergin Shaft. https://mpi.ysn.ru/en/facilities/shergin-shaft. Accessed 26 July 2022.

Nääs, Heta, Noémie Ross, Frédéric Bouchard, Bethany Deshpande, and Michael Fritz. 2017. *Frozen-Ground Cartoons: An International Collaboration Between Artists and Permafrost Scientists*, 27 p. Potsdam: Bibliothek Wissenschaftspark Albert Einstein. https://doi.org/10.2312/GFZ.LIS.2017.001.

Ramage, Justine, Leneisja Jungsberg, Shinan Wang, Sebastian Westermann, Hugues Lantuit, and Timothy Heleniak. 2021. Population Living on Permafrost in the Arctic. *Population and Environment* 43: 22–38. https://doi.org/10.1007/s11111-020-00370-6.

Schuur, E.A.G., A.D. McGuire, C. Schädel, G. Grosse, et al. 2015. Climate Change and the Permafrost Carbon Feedback. *Nature* 520: 171–179. https://doi.org/10.1038/nature14338.

Shiklomanov, Nikolay. 2005. From Exploration to Systematic Investigation: Development of Geocryology in 19th- and Early 20th-Century Russia. *Physical Geography* 26 (4): 249–263. https://doi.org/10.2747/0272-3646.26.4.249.

Tammiksaar, Erki, and Iian R. Stone. 2007. Alexander von Middendorff and His Expedition to Siberia (1842–1845). *Polar Record* 43 (3): 193–216. https://doi.org/10.1017/S0032247407006407.

Van Everdingen, Robert O. 2005. *Multi-Language Glossary of Permafrost and Related Ground-Ice*. Ottawa: International Permafrost Association, Terminology Working Group.

On the Visual Narratives of Ice in Popular Culture: Comics on Ice, Icy Villains and Ice Science

Anna-Sophie Jürgens, Stefan Buchenberger, Laurence Grove, and Matteo Farinella

"Allow me to break the ice": Visual Fictions of Frozen Water and Science in Popular Narratives

"Tonight's forecast: a freeze is coming!" While, for most of us, cold temperatures and snow are simply the weather, in Gotham City you never know. It could be something—someone—else there. It could be Batman's frostiest foe: a former scientist in a cryo-suit who has become a hypothermic ice monster. This monster appears, for example, in Arnold Schwarzenegger's *Batman & Robin* (1997) incarnation, where it produces one bad ice pun after another. For more than 60 years, Mr

A.-S. Jürgens (✉)
Australian National University, Canberra, ACT, Australia
e-mail: anna-sophie.jurgens@anu.edu.au

A. Hemkendreis, A.-S. Jürgens (eds.), *Communicating Ice through Popular Art and Aesthetics*, Palgrave Studies in Media and Environmental Communication,
https://doi.org/10.1007/978-3-031-39787-5_13

Freeze—aka Dr Victor Fries, a cryogenics expert turned cryokinetic (he has the ability to manipulate ice and cold)—has been a pop cultural icon in various media such as (animated) films and comic books. Merging ice and science fantasies, he has shaped our (Western) cultural ideas of both.

In visual fiction and beyond, ice—and melting ice in particular—has become a defining symbol of our collective relationship with the environment, embodying the most existential challenge and threat of the twenty-first century. While scientific data is often seen as the dominant mode of communication regarding anthropogenic climate change, public perception and understandings of our environmental crisis are embedded in a matrix of complex socio-cultural processes that give environmental science meaning in our daily lives (Boykoff and Osnes 2019, 155). Studies of science popularisation demonstrate that "its cultural meanings, and not its knowledge, may be the most significant element contributing to public attitudes toward science" and that popular images of science can "significantly influence public attitudes toward it by shaping, cultivating, or reinforcing these 'cultural meanings' of science" (Kirby 2017, 11). From this perspective, science-related cultural products exploring environmental themes—including comic book stories and animated films—are powerful vehicles of, and for, science and environmental communication.

Visual science fictions—for example in the form of science villains such as Mr Freeze—reflect ideas about science and "construct perceptions for both the public and scientists in a mutual shaping of science and culture" (Kirby 2008, 44). The process behind these phenomena has been described as one "by which the scientific culture and its knowledge become incorporated into the common culture" (Bryant 2003, 357)—which is a media culture. Media culture has been called "the dominant culture today" (Kellner 2020, 15) and "the primary educational force in regulating the meanings, values, and tastes that legitimate particular subject positions" (Giroux 2010, 2). Media culture shapes our daily lives, identity and

S. Buchenberger
Kanagawa University, Yokohama, Japan

L. Grove
University of Glasgow, Glasgow, Scotland

M. Farinella
New York, NY, USA

opinions, "educating us how to behave and what to think, feel, believe, fear, and desire—and what not to" (Kellner 2020, 2). Pop cultural products investigating science and environmental themes—as a form of "public pedagogy" (Giroux 2010, 2)—provide enjoyment and other affective responses and experiences (such as amusement), and can (thus) form, reform or confirm science- and environment-related opinions (Burns et al. 2003, 190; Hee et al. 2022).

Popular media and audience identification with its protagonists are a source of audience influence and affect perceptions of (behavioural) norms and changes (Rhodes and Ellithorpe 2016, 362). In the larger-than-life fictional worlds of icy superheroes—for example, those that are examined below—the ritualised, repeated interactions of familiar characters are a testing ground for how their creators and audiences understand culture, including science. These interactions are useful and interesting because "they provide bold metaphors for discussing ideas or reifying abstractions into narrative fiction" (Wolk 2007, 92). Producing and negotiating meanings, fictional narratives are a powerful means of reaching publics with scientific and environmentalist content and can help audiences interpret information about science (without having to justify its accuracy), which can help them answer questions about the world. Pop culture and narrative (visual) fiction are thus where collective science understandings are created. In other words, cultural meanings and images *of* science can emerge outside the framework of science altogether—and they can "explore and exploit the mirror images of science or scientists in the collective imagination" (Hüppauf and Weingart 2007, 6).

Pop cultural narrative media are increasingly attracting attention as a means of science communication and education. Within this context, comics have been explored as tools for teaching and improving student engagement and motivation (Hosler and Boomer 2011; Jee and Anggoro 2012; Spiegel et al. 2013). Comics can blend engaging character-driven narratives with visual explanations (Farinella 2018a), which are an essential tool in science education (Pauwels 2006). Moreover, comics often use metaphorical representations of the world that "can be mapped onto ours"—and thus they can be more "meaningful in some ways than an accurate depiction of our image world" (Wolk 2007, 134). These visual analogies are central to the public understanding of science (Brown 2003; Farinella 2018b; Gentner and Miall 1981), especially when dealing with abstract, or 'psychologically distant' subjects (McDonald et al. 2015) with which we lack direct experience. However, metaphors often carry complex cultural associations and necessarily distort the target domain (Farinella 2018b), with potentially counterproductive effects (Pauwels 2013).

Against this background, ice-based characters and ice metaphors constitute an ideal case study for visuals' ability to shape—for better or for worse—public perception of science. In this chapter, therefore, we ask: What visual narratives and cultural ideas of ice- and snow-related science do we convey through the medium of sequential art? What do 'comics on ice' and 'ice science' villains teach us about our imagined or unimaginable ecological past, present and future? Exploring these questions in visual fiction—in comics—can deepen our understanding of the intangible cultural aspects of both ice and science (Burns et al. 2003) and open up "novel forms of seeing, of understanding interconnections" (Löschnigg and Braunecker 2019, 4). Ultimately, discussing visual representations of ice, 'comics on ice' and comic book 'ice science' highlights that our cultural and artistic expressions and media "are participants as well as producers of a dialogue about knowledge and have an important function within the public discourse" (Pansegrau 2007, 257). Understanding ice-environmentality and enjoyment incarnated in comics may facilitate a deeper engagement with ecological issues and environmental urgencies, which can contribute to a "healthy scientific culture within society" (Burns et al. 2003, 197).

Comics on Ice

Although there were no comics as such in the Ice Age, visual narratives span human time through what we might call 'proto-comics'—from prehistory to the Ancients, and more recently the Middle Ages onwards. To take but one example, the descriptive ice scenes in Pieter Brueghel's 1565 *Winter Landscape with Ice Skaters and Bird Trap* tell a story: we see a sleepy village, with snow-laden rooves, traversed by a frozen river upon which a couple dozen or so tiny figures indulge in ice games. The viewer can imagine the story that would go along with this image. Is this a comic—or a proto-comic as such? Well, if a single image can contain a narrative, then yes (see Andrews 1995 for a specific reference to pre-modern culture). Here—and this is a point that will underpin the examples to come—the ice and snow themselves are central characters. They are characters that provide polarity as faceless baddies (more on bad guys with faces in the ensuing section), whilst remaining symbolically white—and not black, in terms of traditional binaries—thus providing a nemesis without malevolence, a nemesis without nastiness.

In the early modern period, emblem books often mixed text and image with an implied narrative. Gilles Corrozet's squirrel is one such example (1540).[1] But what has this to do with our theme? The likeness with the poster for *Ice Age 4* is uncanny; the flagship image for this animation is once again a large-tailed squirrel on a raft, this time a block of ice! One might add, in passing, that forms of depiction can travel in different cultures and media and thereby change their contexts: they are thus automobile and powerful.[2] Here the narrative is about life's dangers; indeed, danger and decay are often associated with ice. In the sixteenth century, Geoffrey Whitney's *A Choice of Emblems* includes the story of a fox that takes refuge on ice (Whitney 1586), only for it to melt, illustrating the book's title: *Nullus dolus contra casum*—or "Cunning is nothing against chance". Moving forward, the world's first modern comic, the *Glasgow Looking Glass* of 1825–1826 includes an issue (#15, 23 January 1826) with a frozen pond scene on Duddingston Loch (Fig. 1). This is parallel to Henry Raeburn's 1795 *The Skating Minister*,[3] an oil painting of a

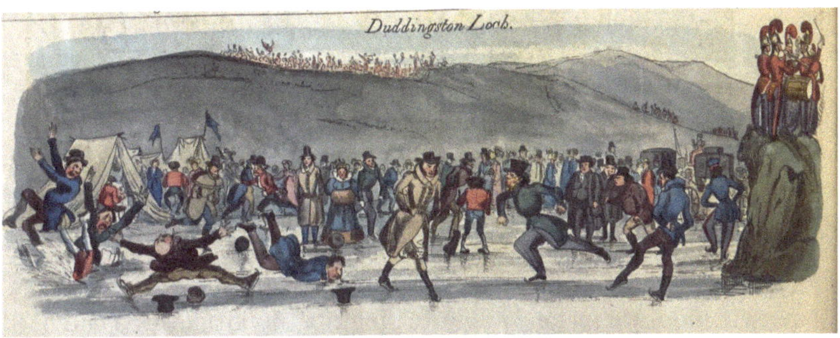

Fig. 1 Duddingston Loch. *Glasgow Looking Glass*, 15, 23 January 1826, © Glasgow University Library, Library (Bh14-x.8)

[1] The illustration is accessible through the Glasgow University Library at https://www.emblems.arts.gla.ac.uk/french/emblem.php?id=FCGa067 (accessed 27 August 2022).

[2] For a contemporary approach to this Warburgian topos, see the research project "Bilderfahrzeuge" (Hypotheses.org).

[3] Sir Henry Raeburn, *Reverend Robert Walker (1755–1808) Skating on Duddingston Loch*, c. 1795. Edinburgh, Scottish National Gallery.

clergyman in full fling on the ice, with all its oppositions of good and evil, black and white, holy and frivolous.[4]

Nearer to what we now call comics, Winsor McCay's *Slumberland* fantasies—the dream world of *Little Nemo*, produced from 1905 to 1911—notably include a Snow King and his realm (McCay 2000), an iconographic precursor, in our view, to *Frozen* (see Chap. 12 in this book). These examples are, of course, the tip of the comic-evolution iceberg, but they serve to show that ice and snow are themes that are present throughout the development of comics narratives, providing oppositions and environments with timeless resonances.

What's Chilling? Exploring Ice Themes in Two Contemporary French Comics: *Période glaciare* and *La Caverne du Pont d'Arc*

As it is time to lace our skates up, let us look specifically at two twenty-first-century snow comics, or *bandes dessinées* (comics in French): *Période glacière* of 2005 (*Glacial Period* in the English version) by Nicolas de Crécy and *La Caverne du Pont de l'Arc* of 2015 by Marc Azéma and Gilles Tosello. Both are contemporary examples of the rich tradition of non-mainstream *bandes dessinées.* In *Glacial Period*, de Crécy creates a futuristic rediscovery of the Louvre and its masterpieces by scientists/future archaeologists after it has been buried in an icy *période glacière.* Soft pastel colours contrast with the predominance of white. A further contrast is created between the notion of art as everlasting, and humans and animals as transitory. Nonetheless, even the everlasting is transformed by the ravages of time; here, symbolised by the new Glacial Period and the concluding scene of icy nothingness. In this work—part of a series commissioned by the Louvre—fiction and metafiction intermingle, as do imagination and lived reality, in a microcosm of the snowstorm that is life itself. Here we see the connection between ice and 'rediscovery', or ice and research, and ice and culture, all in the context of climate change (Fig. 2).

La Caverne du Pont de l'Arc reimagines the struggles of the individuals and animals who lived at the time of the album's setting, around the Pont de l'Arc prehistoric cave in the Ardèche (Fig. 3), south-eastern France.

[4] The image is accessible through the National Galleries Scotland at https://www.nationalgalleries.org/art-and-artists/5327/reverend-robert-walker-1755-1808-skating-duddingston-loch (accessed 27 August 2022).

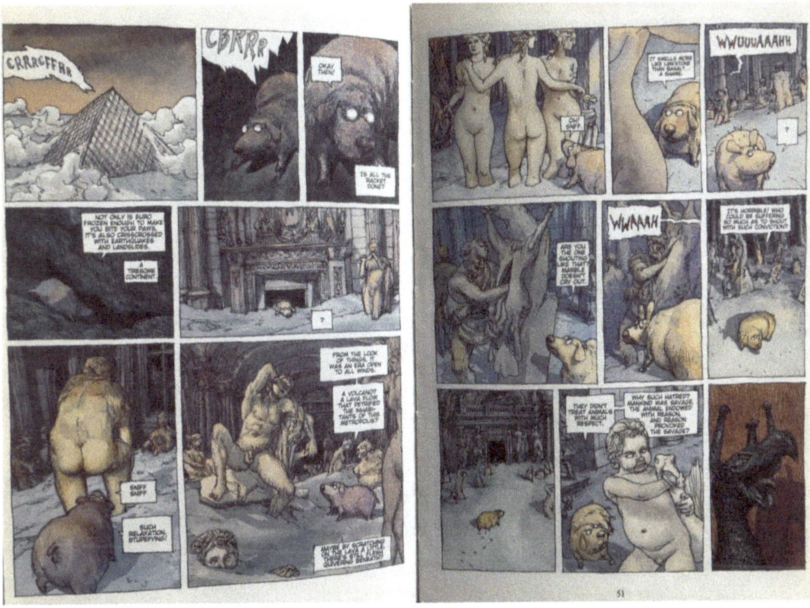

Fig. 2 Nicolas de Crécy, Période glacière, 2005, Paris: Futuropolis/Musée du Louvre, 50–51, in English as *Glacial Period*, New York: NBM Publishing 2006 (with thanks to Florence Briand)

The snow is a historic and anthropological necessity (humans depend on it), but also, again, a representative of humankind's eternal struggle. The image is in a book that takes comic strips into the realm of science, mixing fictional characters from prehistoric times with detailed documentation of related archaeological work—fitting previous aspects of its series—and references that combine art, history, pre-history and anthropology. Here comics mix with related exhibitions and festivals, such as Rencontres d'Archéologie de la Narbonnaise.

Wh(Ice)?

Why Ice? Why does this topic seem so suited to comics, especially given that the aforementioned examples are just a few amongst many? On the one level, it provides opposition. This is an obvious sine qua non for narratives that are often oppositional, but in this case, it is an opposition that

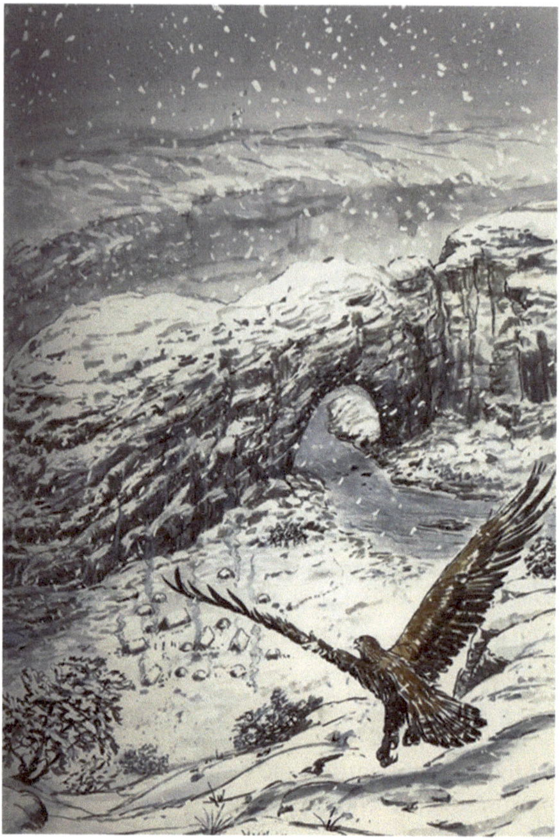

Fig. 3 Marc Azéma and Gilles Tossello. La Caverne du Pont de l'Arc: The Pont d'Arc Cave. Narbonne: Passé Simple, 2015, 31 © Marc Azéma

also allows for assimilation, through the paradox of a friendly enemy. More specifically, we note the eternal nature of storytelling about snow. Indeed, both these forms boast intrinsic hybrid and changing natures, as we have seen from these examples. It is hybrid in that comics mix text and image, and ice is a chemical composition; it is changing in that comics evolve across time and with differing readerships, and ice is often marked by its transience in cycles of freezing and melting. Or more precisely, ice appears as an undefined area between form and the formless, between opposition and the ephemeral, from which imagination can evolve. Science—or, rather, research—in these two ice-related contemporary *bande dessinées* is not about exploring strict scientific methods to understand nature by systematically accumulating scientific 'data', or making risky predictions to follow scientific goals. Yet, the comic book characters do strive for knowledge (about the past, their climate, their world) in order to unlock the secrets of their icy universe. Could they be considered revenants of scientists? After all, the fictional characters in *Période glaciare* and *La Caverne du Pont d'Arc* do explore new frontiers, rebel against the status quo and overall have the ability to see things in a different light, through the eyes of explorers—essential characteristics of scientists (Oliver 1991).

Finally, both comics and snow allow non-visual emotions to be expressed visually: the blast of an icy heart or the Petrarchan contrast of the defrosting of love. What happens to snow and ice in the hands of super villains is another story; ice, snow and the cold themselves are made into a bad guy, which plays into our narratives about the threat of, and to, ice.

Icy Villains in Comics and their (Pseudo-)Ice Science in the DC and Marvel Universes

Superhero comic books are essentially the depiction of a never-ending battle between the forces of good and evil by means of graphic storytelling—or, as comic legend Will Eisner calls it, "sequential art" (2006). As a popular medium, comics also reflect the society they are created by, often leaving a trail of intertextual references to myths, fairy tales, history, politics or science, to name but a few. And since superheroes need convincing adversaries, these contexts create super villains. The cold—which, due to technical advances and global warming, has lost some of the terror it held for our ancestors—is still not to be taken lightly, as annual catastrophic snowstorms or avalanches prove. The historical races to the North and South Pole also provide examples of the struggle against this deadly enemy, which, at its absolute lowest temperature, is the most lethal environment for human beings. Wielding the cold as a weapon, evoking all these forgotten fears and struggles, the ice powers of comic book super villains come in various forms. They originate mythically, are derived from a mutated or alien physique or are the result of comic book (pseudo-)science—where the villains either create their own cold weaponry or are caught in an accident, which gives them their cold powers. Heroes have cold powers too—for example Iceman, a member of Marvel's mutant superhero team X-Men or Superman, who has icy breath to freeze his enemies as one of his many powers—but due to the potential lethality, it is used more often by villains than heroes. However, as will be shown, the most notorious frozen fiends heavily rely on pseudo-science when challenging their heroic counterparts by means of ice and snow.[5] While DC Comics' main frozen fiends almost exclusively rely on (pseudo-)technology and (pseudo-)science and are all, to some degree, the result of accidents, Marvel Comics' most prominent icy villains are primarily from the realm of magic and fantasy, with an added element of science fiction. This also illustrates the antagonistic dynamic between the two major mainstream comic book universes. Spider-Man, for example, has science-based adversaries but no major cold-themed antagonist. Iconic science-based ice villains—like DC's Captain Cold or Mr Freeze, together with their frozen magical counterparts in the Marvel universe—will continue to battle their superhero counterparts weaponising, to various degrees, the cold—the deadliest environment for human beings.[6]

DC Ice Villains

The Icicle—the first major cold-themed mainstream villain, called the Icicle (aka Dr Joar Mahkent)—made his debut in *All American Comics* #90 (1947),[7] where he battled Alan Scott, the first Green Lantern; it was a duel of magic against science fiction-inspired pseudo-science (supposedly based on scientific principles). Green Lantern relies on magic, while the Icicle is a scientist who invents his own super weapons: the cold gun and other cold emitting devices. Originally a villain from the so-called Golden Age of comics, Icicle successfully transitioned into the second period (the 'Silver Age') but was eventually killed in the multi-crossover

[5] In the context of mainstream superhero comic books, science is often used as a form of science fiction. This includes robotics (Marvel's Doctor Octopus) or meteorology (DC's Weather Wizard), fantastical inventions, powers or scenarios, for example time travel (Marvel's Kang the Conqueror), humans running faster than the speed of light (DC's The Flash), or futuristic civilisations that exist in a micro cosmos (Marvel's Microcosmos in Fantastic Four). The sphere of science seems to cover everything that can be used to present the superhero in need to overcome a challenge or a villain.

[6] Images of the following comic book characters can be found online on the CBR platform (e.g. CBR Staff 2017 and Buxton 2016).

[7] For more details and biographical information on all comics referred to in the following, see *The Grand Comics Database* at www.comics.org (accessed 25 July 2022).

event *Crisis on Infinite Earths* (1985–1986).[8] In the post-*Crisis* DC Universe (DCU), the second incarnation of the Icicle is the son of the first one, first appearing in *Infinity Inc.* #34 (1987). His powers also derive from technology, but he does not need a cold gun anymore, since his father's DNA was altered by exposure to his own cold weaponry. Like his father, he is an unrelenting super villain, but has remained a minor character in the DC universe. With the start of the Silver Age, many heroes and villains got redesigned; the Flash, DC's super speedster, arguably started this second age of comics with his first appearance in *Showcase Presents* #4 (1956). Along with his new costume and new secret identity (as police scientist Barry Allen), he also needed new villains to fight. The first major adversary of the second Flash would be DC's second major and, eventually, most notorious, frozen fiend: Captain Cold.

Captain Cold—Captain Cold (aka Lennart Snart) is probably the greatest cold-themed villain in comic books. He first appeared in *Showcase* #8 (1957) in a story aptly called "The Coldest Man on Earth". Another science-based villain, Snart, constructed a cold emitting gun, which accidentally got turned into his iconic cold gun, capable of producing near-absolute zero temperatures (-273.15°C). Contrary to the usual science-based characters in comics, Snart was not a scientist himself, but a blue-collar criminal. His background story was later revamped and enlarged by writer Geoff Johns during his run on the *Flash* (1987-series, #164–225, 2000–2005), giving Captain Cold a kind of definite origin. In

[8] The history of US-American comic books is broadly divided into the following periods or "Ages": The Golden Age (1938–1956) saw the creation of the modern superhero starting with Superman in 1938, followed by Batman in 1939 and other iconic superheroes. However, after WWII superhero comics started to lose their appeal with the growing popularity of crime and horror comics. The comic book industry came under political scrutiny, which led to the industry censoring itself by creating the so-called Comics Code Authority. The Silver Age (1956–1970) arose as a reaction to this when DC introduced the second wave of superheroes and villains, rebooting popular superheroes like The Flash or Green Lantern with new characters and costumes. Marvel Comics started their own line of superhero comics with *Fantastic Four* #1 (1961) and *Amazing Fantasy* #15 (1962), the first appearance of Spider-Man. The Bronze Age (1970–1985) began to address controversial issues such as civil rights, racism, drugs or the Vietnam War. The Modern Age (1985–), also called The Dark Age, represents a trend for more mature comic books which started with two iconic graphic novels: Frank Miller's *Batman: The Dark Knight Returns* (1986) and Alan Moore's *Watchmen* (1986–1987). DC completely rebooted its story universe in *Crisis on Infinite Earths* (1985–1986) transitioning it into the Modern Age, while independent publishers like Dark Horse Comics and Image began to emerge which also led to more creator-owned characters and material (for further information, see Wright 2001).

"Absolute Zero" (*Flash* #182, 2002) Captain Cold's personal history is revealed in an illustrated inner monologue, when he kills his sister's murderer. Snart comes from a broken family, where he and his sister have to live with their abusive alcoholic father. The only relief comes when his grandfather lets the children ride with him in his ice cream truck—where Snart associates the cold with a sense of safety, foreshadowing his future career. He also becomes 'cold' towards people by hiding his emotions, which his father sees as weakness. Caught by the Flash, his heroic archnemesis-to-be, Snart starts reading up on kinetic energy and thermal motion. After his release from prison, he breaks into a science lab and constructs a cold gun from stolen blueprints, which he then powers with cyclotron radiation. However, the cold emitted from his gun is far stronger than Snart had thought, similar to the original Silver Age story. And while DC keeps rebooting its story universe, Johns not only clarified Snart's pseudo-scientific credentials, but also illustrated Snart's connection with the cold, providing him with both comfort and a weapon.

The idea that a common criminal could invent a weapon which accidentally turns into an iconic cold gun does seem to be a bit of a stretch, but when you consider that Flash got his superspeed-powers through a pseudo-scientific accident when he was hit by lightning and subsequently doused in chemicals, it becomes more likely. These kinds of accidental pseudo-scientific origins of superheroes and supervillains were very common in the Silver Age; one only has to think of Spider-Man, who was bitten by a radioactive spider. Captain Cold is, in a way, also the opposite of the speedy Flash, as his gun produces an absolute cold where all movement stops. The villain as the opposite or mirror image of the hero is, of course, a common pattern in antagonistic comic book hero-villain relationships. Captain Cold is not the only member of the Flash's gallery of enemies who use cold-inducing pseudo-scientific weapons: The eponymous Weather Wizard weaponises the cold to create blizzards and hailstorms. Mr Element can control chemical elements with his gun. Captain Cold—together with other members of the Flash's Gallery of Rogues—is also a major character of *The Flash* TV series. In the episode *Snowpack* (*The Flash*, Season 5, 2019) the (aforementioned) Icicle—the evil side of split personality scientist Thomas Snow, who can project extreme cold out of his body—makes an appearance. Here he sacrifices himself to save the life of his daughter Caitlin, aka Killer Frost.

Killer Frost—There are, in fact, three different characters that call themselves Killer Frost in the DC universe, as a literal take on the classic femme fatale. The original, Crystal Frost, first appeared in *Firestorm the Nuclear Man* #3 (1978) and later died in a fight with the titular superhero in *The Fury of Firestorm* #21 (1984). The second Killer Frost, Louise Lincoln, also first appeared in this issue and later duplicated the original experiment that accidentally led to the creation of her predecessor. The current Killer Frost, Caitlin Snow, first appeared as part of DC's *The New 52* reboot in a comic featuring Firestorm (*Fury of Firestorm* #19, 2013)—with whom all Killer Frost incarnations have an obvious antagonistic relationship: a hero whose head is literally on fire versus a villain who is literally frozen. All three derive their cold powers from pseudo-scientific accidents and these powers are, to some extent, based on scientific principles: they absorb heat and turn it into cold. This cold Killer Frost can project blizzards, ice projectiles or freeze matter with her touch.

Mr Freeze—Besides the Flash and Spider-Man, Batman probably has the most prestigious gallery of infamous villains, such as the Joker, the Penguin or Two-Face. One villain with quite a history of fighting Batman is Mr Freeze, who appeared for the first time in 1959 in *Batman* #121 (1959). At first, he was—in the true campy fashion of the Silver Age of comics—called Mr Zero, although he was renamed Mr Freeze in 1966 in the *Batman* TV series. Once again, an accident involving science was the origin of this frozen fiend. Trying to find a cure for his terminally ill wife who had to be kept in a cryogenic capsule, renowned cryogenics expert Victor Fries was caught in a lab explosion that altered his body chemistry; as a result, he had to constantly wear a cryogenic suit, which effectively dehumanised him. Finding a cure for his wife used to be his main (criminal) motivation, but recently he succeeded in curing her. This turned her into a female version of himself—the ruthless criminal Mrs Freeze—in *Detective Comics* #1016 (2020). In a tragic reversal of fortunes, Victor Fries becomes imprisoned in a cryogenic capsule at the end of this story. Here, science does not seem to hold the answer for Mr Freeze, as the problem created by it cannot be solved by it. Cryogenic life extension seems to offer a kind of immortality when human corpses or brains are frozen in the hope that technological advances will someday make it possible to restore them to perfect health. Nora Fries became a vicious criminal upon being released from her cryonic sleep while Mr Freeze will have to wait for his own rescue. However, due to comics' pseudo-science, this will be easier than in real life.

Marvel Comics and Norse Mythology

Of course, Marvel Comics has cold-themed villains too, but most of them are creatures of myth and magic in contrast to DC's pseudo-science villains. This antagonistic dynamic between the two major comic book universes can be seen in many of their respective characters. There is, however, one purely science-based cold villain in the Marvel comic book universe, who was originally named after the mythical Jack Frost, the personification of ice and snow.

Jack Frost/Blizzard—Greg Shapanka was a scientist who first appeared in *Tales of Suspense* #45 (1963) and tried to achieve immortality by means of cryogenics. He developed a cold projecting suit that he used to steal technology from Stark Industries, which made him an enemy of Iron Man aka Tony Stark, the quintessential science hero of the Marvel Comics Universe (MCU). Jack Frost later changed his name to Blizzard and was eventually killed in *Amazing Spider-Man Annual* #20 (1986). He was replaced with a new Blizzard, but this incarnation is also only a minor villain.

Ymir—One source of inspiration for the Marvel founding fathers, Stan Lee and Jack Kirby, was Norse mythology. Marvel's most powerful hero, Thor, not only fought conventional super villains, but many of his main adversaries were mythological creatures: Hel (Goddess of Death), (Fire Demon) Surtr and the Frost Giants, led by their king, Ymir. According to Norse mythology, Ymir was a primaeval being from whose corpse the gods created the Earth. In the Marvel Comic Universe, Ymir is an immensely powerful elemental creature who tries to destroy all life by bringing about eternal winter. Ymir is a source of immense destruction, but offers little narrative complexity, so for all his power he remains a relatively minor villain.

Laufey—Besides Ymir and his frozen minions there is another type of Frost Giants in the MCU, best known from the first *Thor* movie (2011, directed by Kenneth Branagh). Laufey, king of the Giants of Jotunheim, appeared for the first time in *Journey into Mystery* #112 (1965). He was slain by Odin and revealed as the father of Loki, the Norse God of Mischief.

As Loki's father, Laufey is a more complex character than Ymir but still has limited story potential. However, due to the success of the *Thor* movie, he made his comic book comeback more than 50 years later as part of the multi-crossover *War of the Realms* storyline (2018–2019), where he appeared to be styled after the movie villain, attempting to destroy humankind. Thus, instead of the usual adaptation of comics into movies, in Laufey's case there is a reverse influence from movie to comic.

As seen in the series of *Thor* movies and comics (cf. Arnold 2011), the gods of Asgard not only use magic but also science—for example, the rainbow bridge, the Bifrost, that connects Asgard to the rest of the universe, or the Destroyer, Odin's ultimate weapon. *War of the Realms* also saw the re-emergence of Marvel's ultimate cold weapon, the Casket of Ancient Winter. It first appeared in *The Mighty Thor* #345 (1984), where it was used to bring eternal winter to Earth; this is another example of how the magical nature of the cold is used as a force of evil on a way bigger scale in the MCU. It also draws from Norse mythology, as the casket supposedly contains the mythical fimbulwinter, which is the last stage before Ragnarök, the final destruction of the nine realms.

Although the examples from the MCU and DCU represent very different cultural explorations of ice, research in ice and ice research, they go beyond the stereotypical embodiment of science by the ubiquitous scientist in a lab coat. If popular and scientific images of, and about, science create a public imaginary that shapes discourse and attitudes towards science (Bucchi and Saracino 2016; Hüppauf and Weingart 2007), what images and ideas of ice research and ice do our visual narratives actually convey? The icy villains presented here are all manifestations of the inherent natural dangers of ice and cold, and the risks of their abuse for criminal or evil purposes. The villains based on myth and Norse mythology take us back to a time when our ancestors huddled around the fire while snowstorms were howling outside. The villains based on pseudo-science evoke mad scientists from literature—like Victor Frankenstein—and, with it, the dangers of humans trying to control nature; this is an undertaking that often ends in (fictional) disaster. Climate change was created by humans and their science; now humans are using science to try to prevent or at least lessen the catastrophic consequences of their actions. However, the way Captain Cold and his icy companions abuse the cold for their selfish purposes does not inspire much hope.

"Let's kick some ice!"—On the Visual Narratives of Ice and 'Ice Science'

In mainstream comics, ice and snow are associated with rogue ice scientists and weapons, whether they are magical, mythological or 'scientific' in origin. The cold—in the form of ice and snow—is linked to both science experiment and trauma, and is often an incarnation of an ambivalent and ambiguous human struggle. But ice and snow are also central figures beyond the mainstream comics of DC and Marvel, as shown in the aforementioned *bande dessinées* and early visual fictions, in which they are associated with polarity, faceless bodies, danger and decay. Here, the narrative emerges not from what can be killed (or weaponised) by the cold, but from what is *not* hidden in the snow. Snow appears as a strategy of estrangement and entombment, and also as an allegory of the death of an institution (a museum, the Louvre) devoted to the procurement, study and display of objects and knowledge of the past, of lasting interest and value (Flinn 2013, 93). Or, in other words, snow can be the end of art and civilisation. It functions as a mourning veil; it exposes what it conceals.

Images of science and research need not be accurate representations of reality, as "their value is judged by their functionality for resolving a problem, filling gaps in our knowledge, or facilitating knowledge building or transfer" (Pauwels 2006, viii). The current interplay of increasing global temperatures, melting ice caps and rising sea levels may be particularly difficult to grasp for non-scientists. Yet, these are the kind of non-linear complex dynamics that can most benefit from multimodal graphic explanations (Sousanis 2015). Superhero fiction is an effective means of communicating science and science education (see e.g. Gresh and Weinberg 2005). Zehr attributes the viability of superheroes for science communication to a familiar "mental landscape" that allows students to explore unfamiliar concepts in a non-threatening environment (2013). Although depictions in superhero films and comics may help popularise ice science via storytelling (Bergman 2019; Jürgens et al. 2021), it is also important to reflect on what kind of unintended message these characters may bring into popular culture.

In *Introducing Science Through Images*, Maria Gigante reminds us that "making the world of science visible to nonscientists is not the same as making the world of science present and accessible" (2018, 45). In fact, common depictions of scientists which show them immersed in their work—often looking away from us, focused on complicated machinery or surrounded by sterile laboratories—contribute to the perception of science as distant and inaccessible. In that sense, despite their popularity, characters like Mr Freeze, Captain Cold or Laufey may end up reinforcing the stereotype of the 'mad scientist' who is emotionally 'cold' and detached from reality (Jarreau et al. 2019). Ice then can become a metaphor for some of the worse ideas of science.

At the same time—and this underlines the ambivalence inherent in our fictional ice narratives—the humour conveyed throughout the examples explored in this chapter can counteract these adverse aspects. The ridiculously oversized ice cannons and spacesuit-like cryo-costumes of ice villains look hyperbolic, if not clownish, and clearly portray science as a trapdoor towards the unlikely and fantastic. Confusing the Louvre pyramid with an igloo in *Période glacière* (Fig. 2) is funny and adds enjoyment to the reflection on cultural knowledge and research in the fictional future past (which may even negate negative stereotypes of science). Enjoyable science-related pop cultural artefacts provide affective responses and experiences (such as amusement) in relation to the science at stake, which can not only form, reform or confirm science-related opinions (Burns et al., 2003), but also "influence how meanings course through the veins of our social body" (Boykoff and Osnes 2019, 154).

Not surprisingly, perhaps, comparable approaches can be found in the *Frozen-Ground Cartoons* series (see Chap. 12 in this book) and *Antarctic Log* by Karen Romano Young. These comic strips, created *with* scientists, make permafrost science more fun and accessible to the public by giving a voice to ice scientists, reindeer herders and even the animals who are directly affected by the changing climate in the Arctic. Collectively, these stories may help humanise scientists, as well as explain the less obvious effects of melting ice (such as landslides and an increase in mosquito populations), reducing the 'psychological distance' of climate change and recasting ice not as a threat, but as an ally. In other words, comics that frame ice in a relatable way (exploring the comic dimensions of the medium comics) make the reality of ice (and the threat of its loss) more salient and emotion-laden for audiences. Such projects—in which the comic book format acts as a 'boundary spanner' between scientists and diverse audiences (Jonsson and Grafström 2021), visual fiction, storytelling and scientific facts—offer new perspectives about visual language in science communication and ways of communicating and engaging in dialogue about ice research (Igarashi et al. 2020; De Hosson et al. 2018; Wiseman et al. 2021).

So, if we peel back the lid from the scientist's snowgun and take a peek behind Freeze's lair (fittingly, an abandoned ice factory), we might actually discover that our visual ice narratives are not only about extreme environments and outrageous scientific weaponry, but also about icy humour; and that contemporary science-related comic book projects can make us melt for collaborative ice exploration.

Bibliography

Andrews, Lew. 1995. *Story and Space in Renaissance Art: The Rebirth of Continuous Narrative*. Cambridge: Cambridge University Press.

Arnold, Martin. 2011. *Thor: Myth to Marvel*. London: Bloomsbury.

Azéma, Marc, and Gilles Tossello. 2015. *La Caverne du Pont de l'Arc: The Pont d'Arc Cave*. Narbonne: Passé Simple.

Bergman, Daniel. 2019. The "Marvel"-ous Nature of Science: Using Superhero Movies to Teach Methods and Values in Science. *The Science Teacher* 86 (9): 20–25.

Boykoff, Maxwell, and Beth Osnes. 2019. A Laughing Matter? Confronting Climate Change Through Humor. *Political Geography* 68: 154–163. https://doi.org/10.1016/j.polgeo.2018.09.006.

Brown, Theodore L. 2003. *Making Truth: Metaphor in Science*. Chicago: University of Illinois Press.

Bryant, Chris. 2003. Does Australia Need a More Effective Policy of Science Communication? *International Journal for Parasitology* 33 (4): 357–361. https://doi.org/10.1016/s0020-7519(03)00004-3.

Bucchi, Massimiano, and Barbara Saracino. 2016. "Visual Science Literacy": Images and Public Understanding of Science in the Digital Age. *Science Communication* 38 (6): 812–819. https://doi.org/10.1177/1075547016677833.

Burns, Terry, John O'Connor, and Susan Stocklmayer. 2003. Science Communication: A Contemporary Definition. *Public Understanding of Science* 12 (2): 183–202.

Buxton, Marc. 2016. The Coldest Characters in Comic Books. CBR. https://www.cbr.com/the-coldest-characters-in-comic-books/. Accessed 10 Nov 2022.

CBR Staff. 2017. 15 Supervillains with Ice Powers. CBR. https://www.cbr.com/supervillains-with-ice-powers/. Accessed 10 Nov 2022.

Corrozet, Gilles. 1540. *Hecatomgraphie*. Paris: Denis Janot.

De Crécy, Nicolas. 2005. *Période glacière*. Paris: Futuropolis and Musée du Louvre.

De Hosson, Cecile, Laurence Bordenave, Pierre-Laurent Daures, Nicolas Décamp, Christophe Hache, Julie Horoks, Nassima Guediri, and Eirini Matalliotaki-Fouchaux. 2018. Communicating Science Through the Comics & Science

Workshops: The *Sarabandes* Research Project. *Journal of Science Communication* 17 (02): A03:1–20. https://doi.org/10.22323/2.17020203.

Eisner, Will. 2006. *Comics and Sequential Art*. Tamarac: Poorhouse Press.

Farinella, Matteo. 2018a. The Potential of Comics in Science Communication. *Journal of Science Communication* 17 (Y01): 1–17. https://doi.org/10.22323/2.17010401.

———. 2018b. Of Microscopes and Metaphors: Visual Analogy as a Scientific Tool. *The Comics Grid: Journal of Comics Scholarship* 8 (1): 18. https://doi.org/10.16995/cg.130.

Flinn, Margaret C. 2013. High Comics Art: The Louvre and the Bande Dessinée. *European Comic Art* 6 (2): 69–94.

Gaiman, Neil. 2017. *Norse Mythology*. New York: Norton.

Gentner, Dedre, and David Miall. 1981. Are Scientific Analogies Metaphors? In *Metaphor: Problems and Perspectives*, ed. David Miall, 106–132. Brighton: Harvester Press.

Gigante, Maria E. 2018. *Introducing Science through Images: Cases of Visual Popularization*. Columbia: University of South Carolina Press.

Giroux, Henry A. 2010. *The Mouse That Roared: Disney and the End of Innocence*. Lanham: Rowman/Littlefield.

Glasgow Looking Glass / The Northern Looking Glass. Glasgow, 1825–1826. Copies consulted: GUL Bh14-x.8; GUL Bh14-x.10; Mitchell Library 63106299; National Library of Scotland R274b.

Gresh, Lois, and Robert Weinberg. 2005. *The Science of Supervillains*. Hoboken: John Wiley & Sons.

Hee, Matthew, Anna-Sophie Jürgens, Anastasiya Fiadotava, Karina Judd, and Hannah Feldman. 2022. Communicating Urgency Through Humor: School Strike 4 Climate Protest Placards. *Journal of Science Communication* 21 (05): A02:1–26. https://doi.org/10.22323/2.21050202.

Hosler, Jay, and K.B. Boomer. 2011. Are Comic Books an Effective Way to Engage Nonmajors in Learning and Appreciating Science? *CBE Life Sciences Education* 10 (3): 309–317. https://doi.org/10.1187/cbe.10-07-0090.

Hüppauf, Bernd, and Peter Weingart. 2007. Images in and of Science. In *Science Images and Popular Images of the Sciences*, ed. Bernd Hüppauf and Peter Weingart, 3–51. New York: Routledge.

Igarashi, Yasumasa, Nozomi Mizushima, and Hiromi M. Yokoyama. 2020. Manga-Based Risk Communication for the COVID-19 Pandemic: A Case Study of Storytelling That Incorporates a Cultural Context. *Journal of Science Communication* 19 (07): N02, 1–18. https://doi.org/10.22323/2.19070802.

Jarreau, Paige Brown, Imogene A. Cancellare, Becky J. Carmichael, Lance Porter, Daniel Toker, and Samantha Z. Yammine. 2019. Using Selfies to Challenge Public Stereotypes of Scientists. *PloS One* 14 (5): e0216625:1–23. https://doi.org/10.1371/journal.pone.0216625.

Jee, Benjamin D., and Florencia K. Anggoro. 2012. Comic Cognition: Exploring the Potential Cognitive Impacts of Science Comics. *Journal of Cognitive Education and Psychology* 11 (2): 196–208. https://doi.org/10.1891/1945-8959.11.2.196.

Jonsson, Anna, and Maria Grafström. 2021. Rethinking Science Communication: Reflections on What Happens When Science Meets Comic Art. *Journal of Science Communication* 20 (02): Y01, 1–16. https://doi.org/10.22323/2.20020401.

Jürgens, Anna-Sophie, David C. Tscharke, and Jochen Brocks. 2021. From Caligari to Joker: The Clown Prince of Crime's Psychopathic Science. *Journal of Graphic Novels and Comics.* 13 (5): 685–699. https://doi.org/10.1080/21504857.2021.1989005.

Kellner, Douglas. 2020. *Media Culture: Cultural Studies, Identity, and Politics in the Contemporary Moment.* Milton: Taylor & Francis.

Kirby, David. 2008. Cinematic Science. In *Routledge Handbook of Public Communication of Science and Technology*, ed. Massimiano Bucci and Brian Trench, 41–56. London: Routledge.

Kirby, David A. 2017. The Changing Popular Images of Science. In *The Oxford Handbook of the Science of Science Communication*, ed. Hall Jamieson, Dan Kahan, and Dietram A. Scheufele, 1–18. Oxford: Oxford University Press.

Löschnigg, Maria, and Melanie Braunecker. 2019. *Green Matters: Ecocultural Functions of Literature.* Leiden: Brill.

McCay, Winso. 2000. *Little Nemo: 1905–1914.* Ed. Bill Blackbeard. London: Evergreen.

McDonald, Rachel I., Hui Yi Chai, and Ben R. Newell. 2015. Personal Experience and the 'Psychological Distance' of Climate Change: An Integrative Review. *Journal of Environmental Psychology* 44: 109–118. https://doi.org/10.1016/j.jenvp.2015.10.003.

Oliver, Jack. 1991. *The Incomplete Guide to the Art of Discovery.* New York: Columbia University Press.

Pansegrau, Petra. 2007. Stereotypes and Images of Scientists in Fiction Films. In *Science Images and Popular Images of the Sciences*, ed. Bernd Hüppauf and Peter Weingart, 254–263. New York: Taylor & Francis.

Pauwels, Luc, ed. 2006. *Visual Cultures of Science: Rethinking Representational Practices in Knowledge Building and Science Communication.* Hanover: Dartmouth College Press.

Pauwels, Eleonore. 2013. Communication: Mind the Metaphor. Comments and Opinion. *Nature* 500: 523–524. https://doi.org/10.1038/500523a.

Rhodes, Nancy, and Morgan E. Ellithorpe. 2016. Laughing at Risk: Sitcom Laugh Tracks Communicate Norms for Behavior. *Media Psychology* 19 (3): 359–380. https://doi.org/10.1080/15213269.2015.1090908.

Sousanis, Nick. 2015. *Unflattening.* Cambridge: Harvard University Press.

Spiegel, Amy N., Julia McQuillan, Peter Halpin, Camillia Matuk, and Judy Diamond. 2013. Engaging Teenagers with Science Through Comics. *Research in Science Education* 43 (6): 2309–2326. https://doi.org/10.1007/s11165-013-9358-x.

Whitney, Geffrey. 1586. *A Choice of Emblemes.* Leyden: In the House of Christopher Plantin by Francis Raphelengius.

Wiseman, Richard, Jordan Collver, Rik Worth, and Caroline Watt. 2021. Hocus Pocus: Using Comics to Promote Skepticism About the Paranormal. *Journal of Science Communication* 20 (02): A04:1–16. https://doi.org/10.22323/2.20020204.

Wolk, Douglas. 2007. *Reading Comics: How Graphic Novels Work and What They Mean.* New York: Da Capo Press.

Wright, Bradford W. 2001. *Comic Book Nation.* Baltimore/London: Johns Hopkins University Press.

Zehr, E. Paul. 2013. Avengers Assemble! Using Pop-Culture Icons to Communicate Science. *Advances in Physiology Education* 38 (2): 118–123. https://doi.org/10.1152/advan.00146.2013.

Melt for Me: Communicating Ice Empathy Through the Plasticity of Disney

Ben Nickl

Disney on Ice

A subsidiary of The Walt Disney Company, Walt Disney Pictures is a corporation that sells us varied kinds of messages. Most of us may know them for being wrapped in entertaining storylines with singing princesses that cultivate a sense of emotional connectedness between the content and the viewer: from a poisoned, sleeping beauty without agency in *Snow White and the Seven Dwarfs* (1937) to the *Brave* (2012) daughter of a tribal chief who sets out to find her place in society on her own terms, the embedded coloniality in *Moana* (2016) or shifting parental roles in *The Incredibles* (2004).

Despite discrepant views about its impact on society (Dundes 2019, ix; Midkiff and Austin 2021), Disney's global film empire simultaneously mirrors and reshapes contemporary social trends. Each Disney movie

B. Nickl (✉)
University of Sydney, Camperdown, NSW, Australia
e-mail: benjamin.nickl@sydney.edu.au

A. Hemkendreis, A.-S. Jürgens (eds.), *Communicating Ice through Popular Art and Aesthetics*, Palgrave Studies in Media and Environmental Communication,
https://doi.org/10.1007/978-3-031-39787-5_14

presents a coding system that popularises a message and makes it available to millions. Some of these messages deviate from the dominant paradigms of thought, thereby bringing to mainstream culture counter-hegemonic images that engage bourgeois sensibilities (Benjamin 1986). Thinking about the representation of ice in popular entertainment culture, I discuss Disney Pixar's *Frozen* (2013) as just such a counter-hegemonic image; a response to the melting polar icecaps, the climate crisis that has been so extremely visible in numbers, stats and satellite photographs yet absent in globally shared forms of sentiment.

How can the plasmatic ice animation of a Disney film help us experience what we already know about ice, but do not intrinsically feel for it to scale and as 'embodied form of being'?—by which I mean a feeling about the vanishing of ice that billions across the world can share in with focused sense and guided clarity provided by plasmatic feel-view spaces: animated ice creatures, ice powers and ice histories and futures that re-shape ecohuman narratives. In *Frozen*, they all make an appearance when Princess Anna's kingdom is threatened with perpetual winter caused by her older sister Elsa's out-of-control ice magic powers. It is a new appreciation for ice as gift, not curse or threat, and understanding it as both a charming friend and yet still dangerous if wielded like a mindless weapon. Magical snowmen and icy spells that must be broken and healed from the inside, they mould a rebalancing of human-ice-environments and bodies that proves still very much elusive in our off-screen world.

We lose our Arctic Sea ice at a rate of almost 13% per decade, and over the past 30 years, the oldest and thickest Arctic ice has declined by over 95% (Hancock 2022). Should we not feel sad, or at least grieve, all this loss of Arctic ice? Or encounter some sense of anger and annoyance in not just small parts of the population, but all across the planet? But how, and where? Taking the shape of Olaf the magical snowman (amongst other things and beings), *Frozen*'s animated ice interface shows us why we cannot feel in the appropriate dimensions for a large-scale loss of ice—most of it caused by human-induced global warming. For even if we wanted to, an overly general and fuzzily articulated, or as I suggest here, an un-shaped and a non-plasmatic, eco-anxious fear about the loss of Artic ice does not provide us with the necessary forms to channel our feelings about an abstract danger.

But Disney's animators found a way, a multi-dimensional language that makes ice tangible and come to life for both the human eye and heart. This is especially true for animated ice characters like Olaf, who does not

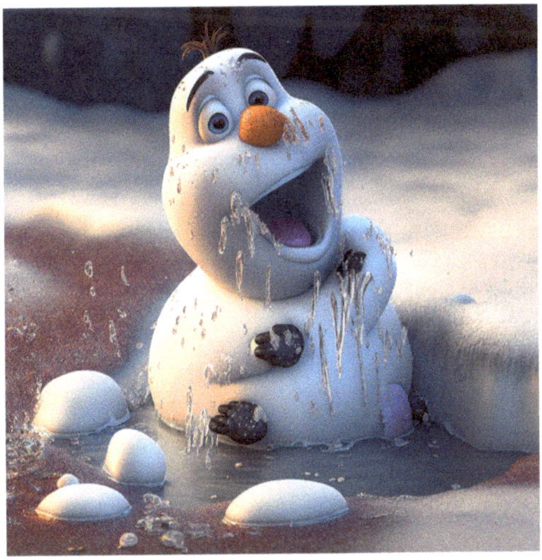

Fig. 1 Benjamin Nickl, Olaf, 2022. © Benjamin Nickl, Midjourney Text-to-Image AI

hesitate to melt and give his life for his best human friend, the princess Anna (Fig. 1)—a process which we can only lament because of the plasmatic quality of Disney's cartoon animation. Olaf's animated ice body offers plasticity of feeling, gives form to our sentiment when it would seem that the beloved snowman is at the brink of death. Here, the plasticity of animated ice provides new shape and imaginative structure and creative contour to the otherwise allotted, ossified idea of ice and its existence for us humans. One might conclude that there is no purposeful feeling without affect-inducing form (Bösel 2021): no empathy for shapeless ice.

This is of particular importance for what is "now being used", as Frank Bierman and others write, "as a conceptual frame by different communities and in a variety of contexts to understand the evolving human-environment relationship" (Biermann et al. 2016, 341): the Anthropocene. A concept widely employed in a variety of contexts, communities and connotations, the notion of an anthropogenic age of human agency and human impact on the planet has attracted the attention of social scientists and humanists. They, and I included here, seek to take the Anthropocene concept beyond its biophysical confines on to discussions that add depth, relevance and scientific value to what some name Earth's latest and perhaps last epoch. A time so shaped by its own looming end, yet seemingly impossible to grasp by human language or to intuit sufficiently by our present capacity of feeling (Anders 1989).

This is the space in which ice humanities emerged in recent years, to refocus from thinking about ice as inanimate volume, minable good and storage, to paying attention to human and non-human actors' intense relationality with it (Leane 2022, 362). For although it symbiotically provides us ecological, geophysical and biochemical services—supplying mobility, water and food security for hundreds of millions of people—ice is primarily the stuff of economics: We—by which I chiefly mean the capitalist West—see it as means for profitable and exploitive resourcing (Dodds and Sörlin 2022, 7–8). What will it take to re-learn and un-know this profit-based extractive logic that we project on ice as something that is there to satisfy only our needs? What does it need from us? Refocusing our affective structures with ice and how we emotionally perceive of it in our daily lives, responding to its varied forms and shapes, may well be one of the first steps along the way (Dodds and Sörlin 2022, 6).

Humanity is after all limited in being able to sense what it means for us and our planet when all ice melts—a horror which the explanatory languages of science and social calls to action have made us less, not more, sensitive to. For such explanatory speech follows styles of communication that replace 'truth' with 'true enough', and deem imperfect understanding on part of audiences as sufficient (Anders 1989): Even though the point was to arrive at truth and not evade it by language that obscures the dire state of things and makes us feel much less instead of more panicked at the prospect of our looming end. In more poignant phrasing, expert scientific warnings of the Anthropocene and polar melting obscure the horror of the end. They exile us emotionally from discourse space we should inhabit all together. Linguistic experting purports to make the public—us—see the danger and the fate that await when sea levels rise and coastal cities drown; what we end up with though are words cladding us in thick-skinned layers of non-feeling that further dissociate us from the truest state of things (Anders 1989).

How true is the end of ice in the way that scientists present it, when they apply a numbing cream to pain and ask us then to act on the masked agony depending on how intense we rate it? And what other tools can science communication use to make us feel the existential threat of climate change—so that we better understand the bond we have with ice and can react accordingly when facing a future that will be without it? For we do not need a language that may move us closer to the facts, but distances us further from the truth at the same time. This is where *Frozen*'s ice animates come in.

Disney's leading ecomagic franchise serves as a case study for the larger question of climate science's relationship with animated storytelling and emotional plasmaticity. The entertainment giant's animation is a kind of communication that can pierce the thick-skinned walls of our hearts, ears and eyes—the totality of an affective assemblage that means 'us'. What is it that we need to see, to hear, to behold and to experience for feelings about ice to become immediately real? Or better yet, what form should that feeling be moulded into—as suggested by the word 'plasma', a term derived from ancient Greek that means 'moulding something into form'? And how can cartoon ice as animated protoplasm take on affect-inducing shapes to stir our capacity to feel? Grounded in a broad reading of ecocriticism in popular entertainment fictions, this set of questions—about the multi-shaped and popularising activity of animation that translates ice into the feel-view language for a post-literary, cinemato-plasmic age—is my focus in this chapter.

The Plasmascope, a Popular Ice-Human Interface

Frozen mediates emotional affinity for different forms and states of frozen water for audiences. One of Disney's highest grossing animated movies of all time, the film helps us form new plasmo-affective relationships with ice (Konnikova 2014). For as it is an elemental protoplasm, ice has no definite form and can assume so many shapes that rouse us into 'feelability' when we encounter anew novel ice things and ice beings. In the interest of focus and space, I group all these relationships under the broader concept of *ice empathy*. It creates—in the most literal of terms—an animated space for human-ice-relationships; shapes for all things ice and bodies frozen-yet-alive. Of course, it would be easy to brush off a Disney sing-along animation musical with catchy tunes and stunning icescapes as pure entertainment with little impact. And indeed, there is no denying that *Frozen*'s ice is fictional. But the emotions that this animated ice evokes, based on plasmatic translation in sensual address, are very real. They may even be more than real to those who grew up with screen entertainment as their window to the world (Müller and Nickl 2021).

It is for this reason, as Ailise Bulfin states, that popular fictions about human-induced change of the environment give us crucial insight into an ecological awareness or structures of feeling that "characterise a society at a particular historic juncture" (2017, 140). At our current juncture, these fictions reflect what millions (if not billions) quite literally see in countless end-of-world blockbuster productions: climate and catastrophe. This further emphasises the need to grasp and to behold these fictions' structures within Disney's massive scale of global pop consumption (Iqbal 2022; Stoll 2022). This is because of, so writes Elia Canetti, the experiential might of many (*mass*): in crowds individuals feel like they are transcending their utter singularity limit (1962, 20). As one of many, the individual stops feeling the inevitable limitation to their own, an isolated body. *We* can do what *I* can not, and *all* of *us* can feel together without distance and act upon that feeling. Disney's entertainment complex forms the most massive kind of such a sensing, collective *we*, spanning across the globe with ease in networks of multiple languages, cultures, ages and communities.

The pop-cultural pedagogy of formable feeling that *Frozen* mobilises through the lens of ice hinges on all of this. Millions of Disneyphiles have watched—and millions more will watch—animated ice that instructs them in its multi-shapedness. It is a cryo-fantasy of sisters and ice creatures that together can beat all the odds and even return both a beloved person and a country from the state of frozen death. The film reenchants audiences collectively anew each time they fall out of faith with the transformative potential of ice, inspiring us to venture past "the ordinary and [onto] extraordinary qualities of this element [to] provide new ways of thinking about ice, living with ice, constructing (with) ice, and ascribing meaning, significance, and value to ice" (Dodds and Sörlin 2022, 3). *Frozen* teaches us how to re-connect to many and varied forms of ice beyond what we assume we know about it—until those feelings reach a critical mass that translates into changed attitudes about ice, moving from the individual to the world stage. Much-needed interventions in climate change discourse can emerge from this mass-cultured emotioneering, as it is done by Disney: an induction of affect, not in a climate of panicked frustration over one's individual futility, but within a collective ice-eco-human thinking channelled as action through the popular might of giant mass (Leane and McGee 2020, 1).

Of course, harnessing the power of 'ice pop' for ice-human interfacing is not new. But as Klaus Dodds and Sverker Sörlin put it in the emerging field of ice humanities, the need to do so with renewed energy could not be greater: the existential crisis posed by the planet's diminishing ice reserves convincingly demonstrates the centrality of ice in human and non-human life (2022). The stakes could not be higher, nor the need for us to emotionally connect to the cryosphere more urgent. There are ever-more changes in climate looming, large and foreboding; we are in a world where glaciers can be unhealthy, "sick or downright dying" (Dodds and Sörlin 2022, 6); and where the experience of snow, frost, winter and cold seasons may become mere fairy tales—for example as in Kevin Reynold's *Waterworld* (1995) and the film's eco-apocalyptic vision of a planet in post-melt catastrophe.

Humanities of ice confront this change with a growing sense of urgency to apply radically different philosophical, social, economic and political views of our environment at various scales. Why not, then, turn to an affective imagining of Arctic ice at the popular level of Disney's global reach? Why not try to feel for ice as for a dying friend, so as to jolt us into expedited climate action to preserve however much we can? With Disney's time-tested technology of three-dimensional feel-view-form animation, within the *plasmascope*—as I will call it here—such jolting is a possibility.

Melt for Me

This brings us to the cryonarrative of *Frozen*. Its plasmatic lens shows ice exists in many ways: as humanoid beings, as monstrous entities, as magical powers and as a fatal curse. Each incarnation elicits distinct emotions; novel ways to think of and feel for ice with an emotional immediacy. We can explore these thoughts and feelings to find out more about who we are. This is because of ice's transmogrification—the many bodies and shapes and forms it can assume—which reflects the knowledge and experience of frozen grounds and waters that have shaped humanity for so many millennia. These bodies and forms shape us as humans just as much as we mould them as non-human beings and things. Ice is intrinsically and undeniably a part of our being in this world, and it has influenced human cultures deep into ancient times and well before. We realise as much while we watch *Frozen*'s shape-shifting states of ice, becoming painstakingly aware that humans live on moving, shifting substances too: earth, rock, ice, water.

These more flexible and malleable ice-world-us relations all become tangible in *Frozen*. With many diverse and different bodies that frozen water can inhabit, the film imagines a range of human feelings for its animated ice. For one, there is overly naïve and child-like Olaf the snowman, a character that we grow to trust, love and care for: warm ice, we may call it. Olaf forms a deep emotional bond with Princess Anna, who is the younger sister of Elsa the Ice Queen. By warming Anna's heart and having his body melt for her (Fig. 1), Olaf melted the hearts of millions of feel-viewers. The melt we felt around the world turns on a human connection to a benevolent, innocent ice creature. Later, this helps Anna melt the metaphorical walls around Elsa's heart by teaching her to not repress emotions (e.g. fear) that stem from her fateful, intrinsic power to create ice.

After pushing her emotions away for so long—specifically the notion that her ice powers are evil and destructive—Elsa does not know how to deal with her feelings in a healthy manner. All she ever knew was to "Conceal, don't feel, don't let them know" as per the lyrics from the iconic *Frozen* track. But how to 'Let It Go' indeed—given the complexity of ice? It can melt to heal us; it can also harden to harm us. This is explained when the film explores the complex dynamics of ice that refuse to "flatten its differences, physical and geopolitical, under the all-encompassing idea of 'The Ice'" (Leane 2022, 361). Elsa tries to turn her emotions off in case she upsets or hurts someone else. She isolates herself because she thinks there is no escape from the icy storm that swirls inside of her. But as she flees her kingdom and her crown, she accidently places ice in Anna's heart. Only true love can break this spell by thawing the malignant ice tumour that grows inside.

Here, ice is coded the most dangerous of things: an aggressor and an insidious infection to the human body. This malicious side of ice also manifests in brutish Marshmallow, a giant, towering, snow monster side character that Elsa creates in a moment of total and absolute fear, to intimidate and attack. All we see of him is a dangerous formulation of ice, beast-like and brutal: roaring, howling, growling. The duality that appears so clearly to feel-viewers at this point in the film is that ice can be cancer, curse and weapon—and a birthing force that spawns all kinds of living ice in many bodies, shapes and forms and states of being. This is how Disney makes its ice all the more intriguing, because the range of things of what plasmatic ice can mean to a human and their environment emerges from within these relations, whether good or bad. So, with each new shape and facet displayed by ice creatures and ice powers, there is 'no one way' for ice to be—as any claim to static homogeneity immediately undoes itself in *Frozen*'s narrative arc.

Compressed further in the apparently bipolar attributes of animated ice are multiple meanings, memories and values. They compete with each other at times, but they never lose sight of the overwhelming concern that binds them all together: the challenge of sustainable futures for ice and all kinds of life, together in symbiotic harmony, whether set in a world of global warming or of cooling. *Frozen*'s overarching ice narrative tells of a whole ecosphere that slowly turns into an icy wasteland, freezing to death people, gnomes, fairies, trolls, animals and plants alike. Emotionally unbalanced, Elsa's distress is a harbinger of just how important a steady emotional relationship with ice is to sustain the environmental conditions for human and non-human life. She has to master her magical ice powers to avert a looming disaster that threatens all things living and thriving in the fictional kingdom of Arendelle. All the while, Anna must learn to love through the cold to melt away the emotional distance to Elsa—navigating, for example, the older sister's maze of an ice palace or pushing ever closer to her sibling through the threatening winds of the snow queen's magic snowstorms.

This moving introspective—triggered by ice in *Frozen*'s narrative—is a catalyst for innovative appreciations: of not a singular substance of ice, but many kinds of ices, as pointed to by Leslie Carol Roberts (2022). Ice empathy, as made experienceable (in the Latin sense of what it means to animate by 'giving life to') through the Disney plasmascope, means emotional affinity for all things frozen in ourselves and in the lives we lead with others: hearts, feelings, minds and barriers to futures yet to melt so we can pass through them and move on. Ice can warm us, like Olaf; kill us, like it does to Anna's body from the inside out; heal us, as it does heal her eventually; hurt us; change us. All this prompts feel-viewers to rethink ice and their feelings for it with a new and complicated idea of what the coexistence with it means—*must* mean.

The changing shapes, functions and forms of ice allow us to understand, to formulate pivotal inquiry. Is ice mere resource and a mindless tool? A threat to our lives? Or is it a friend and voice that we should listen to? The recurrent premise of *Frozen* is that these questions—and others about one's identity and place in the world in relation to ice—are not easy to resolve. Answers only come from critical affective agency, from the engagement with climate bodies that our actions put in crisis in the first place, as so asserted with great insight by Janine Randerson (Randerson 2018). For it is love of self, love of Other and love of ice deeply embedded in the human soul, that saves Anna from cold death. True love, as Olaf describes this feeling, melts Anna's frozen heart and allows Elsa to undo the freeze spell put on Arendelle. Shown as a plasmo-affective reaction when adding love to a crucible of dead ice, both the body-turned-ice-statue of Anna and the ice-covered lands revert back to their un-frozen states. Lifeless ice turns into human flesh once more, and the tundra that was Arendelle melts back into a wondrous place of spring when Elsa is able to control her ice magic.

Thus, the ice-human-feeling balance is restored, and the story of ice empathy in *Frozen* ends when humans re-learn a healthy and respectful relationship with ice, and reverse its catastrophic effects on ecologies and bodies alike. It is a plasmatic, bendy sort of ice that teaches humans to be more than just one thing—to mould and ply their inner form just as animated ice so masterfully does. If they do not learn this, the hurtful, out-of-balance relationships between ice and humans will not resolve in happy endings for either of the two. Michael Richardson suggests that pained human-ecology relationships in ecofictions such as *Frozen* can be read as "jarring, rupturing, disjunctive experiences of future [climate] crisis in the now. [...] Already arriving from the future yet only just beginning to unfold" (2018, 1). Feel-viewers of the *Frozen II* sequel, where Elsa along with Anna and Olaf must yet again establish balance between environment and human interference, will certainly know of ice as a climate crisis 'hyperobject' (Morton 2013; Richardson 2018, 2)—an object whose meanings intrude on the present from the future as first so defined by Timothy Morton (Morton, 2013). But already in the first *Frozen* film and its plasmo-affective aesthetics, ice throws into relief the challenges of conceptualising, experiencing and responding to a highly complex relationship between humans and ice.

The larger project that the story of *Frozen* points to is hence rendering us fluent in this new relationality with climate, to motivate reconsideration of ice in the context of collective symbolism and as a symbol of the very real collective that is us and all other things that we co-exist with. The Disney narrative here is embedded in wondrous literatures of ice discovery and ice encounter, as Andreas Homann shows by documenting literature's long line of scientific ice representations and fictional stories about the cold that combine popular and critical cultural science (2017, 776). In this wider context, animated ice emerges in service of enviro-affective literacy (Brereton 2018). As animated matter, ice has a materiality of emotion that oscillates between a solid substance and psychosocial senses. In that sense, *Frozen*'s animators code ice not only as a physical substance, but also—and, in fact, primarily—as formable feelings at the end of being human. It is this end that leads us one last time into the animated experience in Disney's eco-cultural imagination, displaying trans-species and trans-affective metamorphosis where ice is born from our love and where it does not hesitate to die so we may live.

Animating Olaf

The transformative potential of ice and the emotions that its avatars generate are the primary drivers of *Frozen*'s plot, although the film's ice empathy owes much of its popularity to its animated medium. It is animation that gives body and shape to the cold and icy existences in the world of Arendelle; a magical plasticity of feeling that comes in all kinds of bendable and pliant bodies. This is primarily seen in Olaf the snowman, who springs from Elsa's magical ice powers in a miraculous act of creation that Lauren Dundes et al. attribute to a postfeminist quality in *Frozen*'s psychosocial design (2019). The film's narrative hinges on its two singing Disney princesses' quest to prevent their world from permanent frost, to let Elsa discover her inner balance and to heal Anna's ice curse. It is a high-stakes premise—one that offers a narrative of anthropogenic salvation to feel-viewers who watch and feel along as all the (sub)plot(s) do unfold. In all this, Olaf the heartwarming ice animate proves inherently essential (Fig. 2).

Olaf is not the only plasmatic interface that connects ice with humans in *Frozen*'s plot, but it would be impossible to engage in the affective journey without the bendy-bodied snowman. It is his animatedness that demonstrates what Disney's plasmascope can shape into life-like experience for us: fun, elation, fear and even startled moments of upset and grievous courage that make Olaf melt and then re-form and re-solidify. Olaf's body of feelable ice stretches into the realm of human emotion, or whatever it is that humans can sense and name as such. Beholding this rubber-hose-styled stretching provides the feel-viewer with a mould for perceiving ice that is ironically an anti-mould. When Olaf clowns about with a drooping face or finds a mid-section of his rotund body suddenly set atop his head, he points to the perpetual condition of animated possibility. His cartoon body can *be* in ways, and *feel* in ways, that inter-'face' between a human world of static form and the spectacle of unbound (icy) primordial emotion.

A human only has the static anatomy of their face to do all this, which does not come close to the omni-plasmaticness modelled by every single part of the snowman's physical shape. It is Olaf's constant flux in feel-form that impresses deeply on humans who witness such affectively coded ice plasma. He presents us with a 'matter' centred on cinematic traditions that filmmaking has always drawn upon—and in the case of Disney, literally drawn on: "physiognomy, caricature, commedia dell'arte" (Geil 2019,

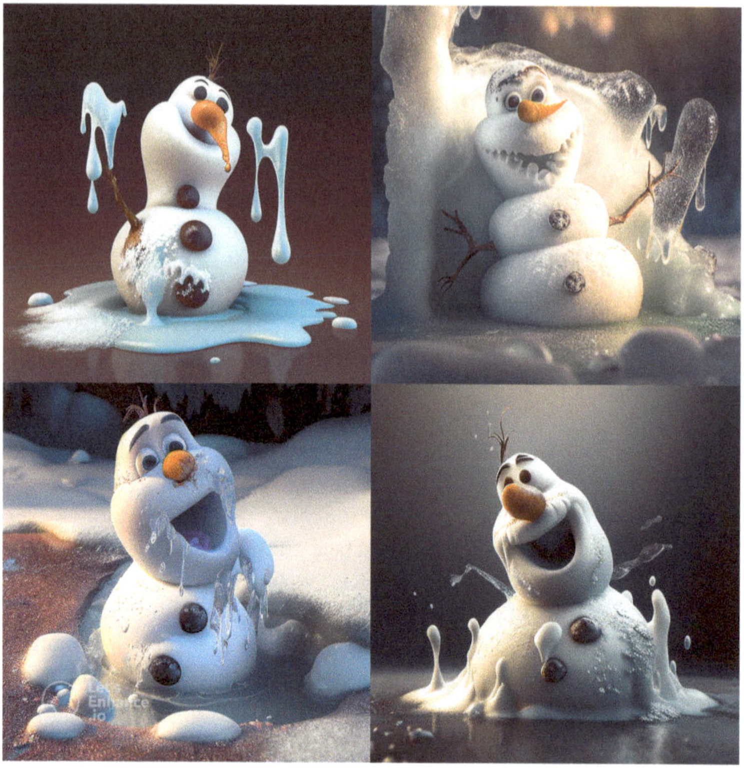

Fig. 2 Benjamin Nickl, Olaf, 2022. © Benjamin Nickl, Midjourney Text-to-Image AI

28). All this makes spectators ask: what if we could do the same? What if we could be in body just as unrestrained and unrestricted in depth of feeling, as Olaf seems to be? A dream of plasmatic ice mimesis, as Abraham Geil betitles this affective aesthetics of animation. It undergirds the feel-view mechanics that the Walt Disney Company has developed with billions of dollars for nearly now a century (Geil 2019, 25; Thomas and Johnston, 1990).

It is here that plasmatic animation reveals the pure power of experiencing primal ice formability—an almost ecstatic sensing of the ability to live as ice while split from human form. "In a word", writes Sergei Eisenstein, the animated form "it is everything that forces the viewer to 'be beside himself'" (1986, 27), that is to be more ice than human in the case of *Frozen*. Therefore, in Olaf's flexible form, we vividly see how plasmatic (non-human) feeling not only strains the limits of bodily human anatomy, but also tests the very ability to feel beyond the limits of the human. It is the snowman's animatedness that produces a gateway to a transhuman affect of the cold element. Via an ice-body interface—that, to some degree, also anthropomorphs frozen water—Olaf endows ice with an irresistible changeability and malleable mode of existence (Väliaho 2017). This is the plasmatic quality and particular aesthetic interest of Disney animation that Eisenstein as famed director of experimental film established in the early 1940s; he argued that animation holds unique currency by shaping worlds and characters to suit ideas and express feelings by rejecting the constraint of (human) form (1986).

The shape-shifting qualities of ice demand elasticity from us as we behold them. Disney puts these plasmatic impulses on screen with the ingenious ice avatar of Olaf. Watching him simply compels us to bend our imagination of hyperhuman ice along with all the twists and turns that Olaf's frosty body can achieve; protoplasmic, 'avant tout'. This revolves around an ecoemotional pedagogy of formable feeling that Disney inserted into the DNA of its so-called magic kingdom of plasmatic animation. Walter Elias Disney and his army of animators built their brand on the dream—or, rather, the industrialised illusion—of bringing fantasies to life through feel-forms, embodied in colour and other pictorial dimensions. It is seen in the many-metred forms that Mickey Mouse's arms can stretch, then again snap back, into like rubber bands; transformance of a body at the animator's whim. Such is the psycho-technological hallmark of Disney's cinema of tangible fantasy (Watts 1995, 84–85): the rejection of, and liberation from, the once-and-forever allotted form of being human.

The conceptual basis for all this rests on the allure of multi-bodied feeling and a shared social attraction to experiencing animated ice plasma as both "profound in thought and irresistibly attractive and exciting in form" (Eisenstein 1986, 101); which all harks back to the initial idea here that the Walt Disney animation studios can mass produce an experience of ice and purposefully make it embody an affective meaning (Bösel 2021): synthetic feelings for synthetically shaped ice. We have seen the mediated mobilisation of such feeling, or engineered emotion, for environmental issues and scarce resources at work before in Disney. The animated blockbuster hit *WALL·E* (2008) can be analysed through the lens of eco-entertainment as done by David Whitley (Whitley 2014), and discourses of environmental harm in *Dr Seuss' The Lorax* (2012) can be critically assessed like Ellen Moore (Moore 2017) has shown. What I have discussed here—*Frozen* as *ice empathy*—is part of that very same tech of plasmo-affective emotioneering that all these Disney animation films turn on in a bid to 'disneyfy' climate action (Midkiff and Austin 2021) and our emotional responsibility to engage with ice close up through feeling and not dwell in distant, hollow words (Brereton 2014).

By pointing this out—and by suggesting an effective nexus of science communication, Ice Humanities and animation as feel-viewing in the age of the Big Melt—I hope to contribute a popular plasmatic approach to science stories and science communication. "By mobilising", as outlined by Sarah Rachael Davies and others, "ideas about experience; by framing science communication through identity work; by focusing on fiction; and by paying attention to emotion" (Davies et al. 2019, 1). All as extensively discussed elsewhere in David Kirby's scholarship on the link between science and popular culture and, especially, audio-visual media (Kirby 2003, 2017).

Bibliography

Anders, Günther. 1989. Sprache und Endzeit (II). *FORVM*. http://forvm.contextxxi.org/sprache-und-endzeit-ii.html. Accessed 24 September 2022.

Benjamin, Walter. 1986. The Work of Art in the Age of Mechanical Reproduction. In *Illuminations*, ed. Hannah Arendt, 217–252. New York: Schocken Books.

Biermann, Frank, Xuemei Bai, Ninad Bondre, Wendy Broadgate, Chen-Tung Arthur Chen, Opha Pauline Dube, Jan Willem Erisman, Marion Glaser, Sandra van der Hel, Maria Carmen Lemos, Sybil Seitzinger, and Karen C. Seto. 2016. Down to Earth: Contextualizing the Anthropocene. *Global Environmental Change* 39: 341–350. https://doi.org/10.1016/j.gloenvcha.2015.11.004.

Bösel, Bernd. 2021. *Die Plastizität der Gefühle: Das affective Leben zwischen Psychotechnik und Ereignis*. Frankfurt: Campus.

Brereton, Pat. 2014. Animated Ecocinema and affect. In *Moving Environments: Affect, Emotion, Ecology, and Film*, ed. Alexa Weik von Mossner, 181–200. Waterloo: Wilfried Laurier University Press.

———. 2018. *Environmental Literacy and New Digital Audiences*. London: Routledge.

Buck, Chris, and Jennifer Lee. 2013. *Frozen*, Wall Disney Pictures, 101 minutes.

———. 2019. *Frozen II*, Wall Disney Pictures, 103 minutes.

Bulfin, Ailise. 2017. Popular Culture and the 'New Human Condition': Catastrophe Narratives and Climate Change. *Global and Planetary Change* 156: 140–146. https://doi.org/10.1016/j.gloplacha.2017.03.002.

Canetti, Elias. 1962. *Crowds and Power*. Trans. Carol Stewart. New York: The Viking Press.

Davies, Sarah Rachael, Megan Halpern, Maja Horst, David Kirby, and Bruce Lewenstein. 2019. Science Stories as Culture: Experience, Identity, Narrative and Emotion in Public Communication of Science. *Journal of Science Communication* 18 (5): A01:1–7. https://doi.org/10.22323/2.18050201.

Dodds, Klaus, and Sverker Sörlin. 2022. Introduction. In *Ice Humanities: Living, Working and Thinking in a Melting World*, ed. Klaus Dodds and Sverker Sörlin, 1–34. Manchester: Manchester University Press.

Dundes, Lauren. 2019. Preface to 'The Psychosocial Implications of Disney Movies'. *Social Sciences* 7: ix–xii.

Dundes, Lauren, Madeline Streiff, and Zachary Streiff. 2019. Storm Power, an Icy Tower and Elsa's Bower: The Winds of Change in Disney's *Frozen*. *Social Sciences* 7: 202–230. https://doi.org/10.3390/socsci7060086.

Eisenstein, Sergei. 1986. *Eisenstein on Disney*. Trans. Alan Upchurch. Kolkata: Seagull Books.

Geil, Abraham. 2019. Plasmatic Mimesis: Notes on Eisenstein's (inter)faces. *World Literature Studies* 11: 26–41.

Hancock, Lorin. 2022. Six Ways Loss of Arctic ice Impacts Everyone. World Wildlife Fund. https://www.worldwildlife.org/pages/six-ways-loss-of-arctic-ice-impacts-everyone. Accessed 12 Sept 2022.

Homann, Andreas. 2017. *Am Ende der Eiszeit*. Leiden: Brill.

Iqbal, Mansoor. 2022. Disney Plus Revenue and Usage Statistics. Business of Apps. https://www.businessofapps.com/data/disney-plus-statistics/. Accessed 2 Aug 2022.

Kirby, David. 2003. Scientists on the Set: Science Consultants and the Communication of Science in Visual Fiction. *Public Understanding of Science* 12: 261–278. https://doi.org/10.1177/0963662503123005.

———. 2017. The Changing Popular Images of Science. In *The Changing Popular Images of Science*, ed. Kathleen Hall Jamieson, Dan Kahan, and Dietram Scheufele, 291–300. London: Oxford University Press.

Konnikova, Maria. 2014. How '*Frozen*' Took over the World. *The New Yorker*. https://www.newyorker.com/science/maria-konnikova/how-frozen-took-over-the-world. Accessed 12 Sept 2022.

Leane, Elizabeth. 2022. Ice Islands of the Anthropocene: The Cultural Meanings of Antarctic Bergs. In *Postcolonial Literatures of Climate Change*, ed. John Ryan, Pauline Reynolds, and Russell McDougall, 361–382. Leiden: Brill.

Leane, Elizabeth, and Jeffrey McGee. 2020. Anthropocene Antarctica: Approaches, Issues and Debates. In *Anthropocene Antarctica: Perspectives from the Humanities, Law, and Social Sciences*, ed. Elizabeth Leane and Jeffrey McGee, 1–14. New York: Routledge.

Midkiff, Emily, and Sara Austin. 2021. The Disneyfication of Climate Crisis: Negotiating Responsibility and Climate Action in *Frozen, Moana*, and *Frozen 2*. *The Lion and the Unicorn* 45: 154–171. https://doi.org/10.1353/uni.2021.0013.
Moore, Ellen. 2017. *Landscape and the Environment in Hollywood Film*. Cham: Palgrave Macmillan.
Morton, Timothy. 2013. *Hyperobjects: Philosophy and Ecology After the End of the World*. Minneapolis: University of Minnesota Press.
Müller, Christopher John, and Benjamin Nickl. 2021. Real is Not Real Enough—Sonic Translation. Goethe Institute Global. https://www.goethe.de/ins/au/en/kul/lok/gap/sot.html. Accessed 1 Aug 2022.
Randerson, Janine. 2018. *Weather as Medium: Toward a Meteorological Art*. Cambridge, MA: MIT Press.
Renaud, Chris, and Kyle Balda. 2012. *Dr. Seuss' The Lorax*, Janet Healy, 86 minutes.
Reynolds, Kevin, and Kevin Costner. 1995. *Waterworld*, Kevin Costner, 130 minutes.
Richardson, Michael. 2018. Climate Trauma, or the Affects of the Catastrophe to Come. *Environmental Humanities* 10: 1–19. https://doi.org/10.1215/22011919-4385444.
Roberts, Leslie Carol. 2022. The Gigaton Ice Theatre: Performing Ecoactivism in Antarctica. In *Performing Ice*, ed. Carolyn Philpott, Elizabeth Leane, and Matt Delbridge, 195–213. Cham: Palgrave Macmillan.
Stanton, Andrew. 2008. *WALL·E*, Morris Jim, 98 minutes.
Stoll, Julia. 2022. Global Number of Disney+ Subscribers 2022. Statista. https://www.statista.com/statistics/1095372/disney-plus-number-of-subscribers-us/. Accessed 5 July 2022.
Thomas, Frank, and Ollie Johnston. 1990. *Disney Animation: The Illusion of Life*. New York: Abbeville Press.
Väliaho, Pasi. 2017. Animation and the Powers of Plasticity. *Animation* 12: 259–271. https://doi.org/10.1177/1746847717740093.
Watts, Steven. 1995. Walt Disney: Art and Politics in the American Century. *Journal of American History* 82: 84–110. https://doi.org/10.2307/2081916.
Whitley, David. 2014. The Wild and the Cute: Disney Animation and Environmental Awareness. In *Kidding Around: The Child in Film and Media*, ed. Alexander Howe and Wynn Yarbrough, 211–222. London: Bloomsbury.

On the Aesthetic Facets of Ice Urgency: Some Final Reflections

Anne Hemkendreis, Anna-Sophie Jürgens, and Karina Judd

A reflecting lake forms the stage for an icy forest landscape, which divides the photograph into an above and a below: into a relatively barren sky through which isolated wisps of clouds drift and into a miraculous lake that is completely still and smooth. As a silent stage of ice—without a trace of life—the lake appears sublime and timeless, like the icy landscape in the distance. However, the impression of a polar landscape with rapturous distance in Konrad Lenz's infrared photograph is just as misleading as its palpable cold. In truth, it is the urban Lake Burley Griffin in Canberra, Australia, which was shimmering in the summer heat at the time the picture was taken. Lenz works with alienating photographic techniques to

A. Hemkendreis (✉)
University of Freiburg, Freiburg im Breisgau, Germany
e-mail: anne.hemkendreis@sfb948.uni-freiburg.de

A.-S. Jürgens • K. Judd
Australian National University, Canberra, ACT, Australia

A. Hemkendreis, A.-S. Jürgens (eds.), *Communicating Ice through Popular Art and Aesthetics*, Palgrave Studies in Media and Environmental Communication,
https://doi.org/10.1007/978-3-031-39787-5_15

question the limits of our perception, our idea of temporal processes and the scope of our imagination. His camera captures a spectrum of light that remains invisible to our everyday perception; the framing of our view of nature—its stereotypes and medial shaping—is thrown into question. This is one of the core themes of this edited volume, explored by its many authors from different angles: the power of ice to stimulate our fantasy, the different ways it can (not) be visualised, the various dimensions in which it has inspired our creativity across media. Ice is invisible, hyper-visible, a mirror and a looking glass. It teaches us as much about itself as it does about us. Throughout this volume, whether ice is discussed as mythical and epical, representational, metaphorical or speculative, eco-emotional or geopolitical, it is a powerful symbol of our collective relationship with the environment (Fig. 1).

Icy landscapes have a firm place in our collective memory. Since ice has been part of visual arts, it has served as a stage for human catastrophes and/or social pleasures. Today, ice and snow can no longer be viewed with naivety; they question invisible dynamics and changes in nature for which industrialised countries are responsible. Whereas artists of the

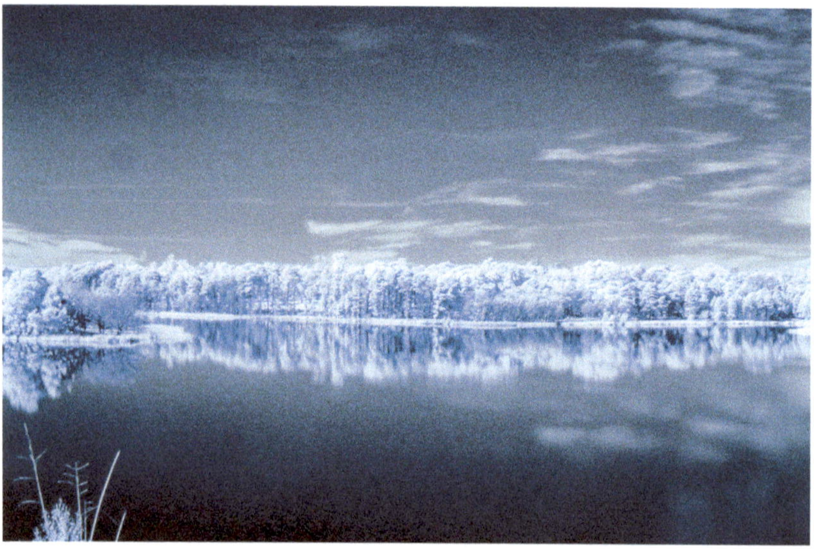

Fig. 1 Konrad Lenz, *Lake Like Ice*. Infrared Photograph, 2018 © Konrad Lenz

industrialisation period started to notice the human impact on landscapes and developed aesthetic strategies to communicate the feeling of loss, today artists are faced with the fact that almost every area of human interest, activity and endeavour is about to be, or already is, impacted by climate change. Melting ice on the poles and basically all continents is one of its most shocking expressions. However, as a catastrophe without a clear beginning or one dramatic precipitating event, climate change has a fundamental problem of scaling, visualisation and, thus, mediation, which makes it difficult to actually communicate and address. This is where art and (popular) culture come into play, creating intense images, stories and memories that stick with the recipient while making scientific knowledge and research around climate change, environmental fragility and the need to act more tangible and accessible. They give meaning and context to data, figures and models from scientific research and help foster an emotional response. However, the 'emotionalisation of (climate) knowledge' also leaves space for self-critical reflections, especially if romantic traditions are quoted and questioned at the very same time.

Images of polar landscapes in the media, for example, suggest an encounter with the effects of our actions. This is achieved by evoking the feeling of familiarity based on the firm anchoring of Polar landscapes in our visual memory. But how many people have had the privilege of travelling to the remote regions—North or South? Or should tourism to the (Ant-)Arctic regions only be imaginary, so as not to contribute to ruining these treasured regions with fuels, diesel, rubbish etc.? The reception of the infrared photograph by Australian artist Konrad Lenz raises these, and other questions. Lenz's *Lake Like Ice* negotiates the limits of our perception, the longing for an encounter with an intact nature and the possibilities of imparting knowledge through aesthetic means. In using sublime aesthetics and showing seemingly vast and unlimited ice regions, Lenz plays with the way in which ice regions are perceived depending on our visual cultural memory. With the 'technique of icing' (using an infrared camera), the widespread interpretation of photography as a medium of the documentary is doubted: The view onto a polar scene is unmasked as an urban landscape. Interpreted boldly, Konrad Lenz's infrared photographs use ice as an aesthetic tool to visualise our cultural imaginaries connected to ice, our relation to it as an (un)familiar element and its ongoing ecological loss.

Communication, Art and Action: Behavioural Change Through (Experienced) Aesthetics?

How can we capture the slippery part of ice and make sense of it—as embodied, for example, in 'Antarctic' plays that are not set in Antarctica? What exactly does ice communicate? And how can we capture the different understandings and views of 'communication' explored in art, popular art and in this interdisciplinary edited volume? The authors in this book do not understand the term 'communication' as a one-dimensional, scientific, social or cultural message to be conveyed. Artists relating to scientific knowledge or scientists collaborating with artists and pop cultural phenomena responding to scientific appeals always may stress urgency, but they also allow the recipients to find their own way to respond. Science-related art may be emotionally moving, but at the same time it opens up room for negotiation and reflection. This is important to bear in mind, to thwart assumptions that there is a linear path from an environmental message to environmental action via an aesthetic experience.

As a material with agency, ice provides room for critical reflection by creating an aesthetic encounter with the other: the non-human. More so, ice art opens up a space of sensual experience in which ice appears as an active agent of knowledge production and reflection (Leane et al. 2020, 14). However, ice also has a community-building effect. The aesthetic experience of ice—for example in an exhibition or cinema context—can lead to an 'affective arrangement', meaning the creation of a network of recipients who are related to each other by a shared aesthetic experience and mutual affection of emotional responses (Slaby and von Scheve 2019). In our view, these types of networks play a key role in (environmental) action as they describe how an intense experience can turn into a shared feeling of urgency. As we argue, these network groups share mutual convictions or intentions, like the urge to act in the face of climate change. This, then, is how a relationship with art related to ice sciences brings about behavioural change: by forming and activating networks among recipients who are emotionally affected and thus more likely to act according to a mutual (here environmental) purpose. However, the question of how exactly affects move within these specific networks of recipients requires further research from different disciplines.

Part one of this volume examined ice on the stage—and ice as the stage itself. In this way, ice acts interchangeably as the medium, the message and the provocation for audiences to make their own meaning. The extended reflexive approach of Graham J. Walker in chapter "Here Be Science Show Dragons: Ice, Icons and Metaphoric Approaches to Climate Change Communication" (this volume) to consider the positionality of both himself and his audiences and evolve his practice provides insight into how both subtle and major shifts in context—including temporal, geographic and cultural—require different approaches. This point echoes throughout this book, where, for example, Dodds (chapter "Ice Stages and Staging Ice") and Schneider (chapter "Materiality of Time: Polar Ice as a Medium for Ecological Art for the Tempered Zones") question how the melting ice presented at Glasgow's COP26 would have been received at previous COP meetings, and whether or not a work like this would have changed any outcomes.

From a Science Communication perspective, the arts have been described as (becoming) a preferred medium for conveying science to publics because they are visually stimulating (Schwartz 2014), spur action, for example on environmental issues, and deepen engagement (Lesen et al. 2016). Research on the power of art in climate change discourses highlights its ability to foster agency and inspire feelings of hope, responsibility and care to create openness beyond the human world and to raise awareness and creativity for tackling complex environmental problems (Bentz 2020). Art also enhances learning through more creative and richer intellectual inquiry when explored in and for science and environmental communication projects (Bentz 2020; Lesen et al. 2016; Warner 2022; Zaelzer 2020). Furthermore, art can be used as an unthreatening approach to science engagement in new spaces and communities where access to scientific literacy has historically been restricted (Streicher et al. 2014). It is a medium which is able, in some cases, to engage 'new' audiences outside the universities, museums and institutions where science 'lives' (Bisbee O'Connell et al. 2020). However, Science Communication literature (Featherstone 2014) does also warn that art is not a panacea for science engagement—while it increases access for some, it alienates others.

As Campos and Araújo (2017) argue, Indigenous artistic expressions are an underutilised and under-researched area in the art-science interface, and the same holds true of this volume on icy environments. Future work should seek to involve—not just consult or research—Indigenous peoples through participatory research (see e.g. Kuhn and Muller 1993; Quick and Feldman 2011) to better understand existing practices and the challenges and opportunities for new work, as Sjöberg et al. (2019) demonstrate. As some authors in this book note, Indigenous peoples in icy environments (and indeed in other types of environments around the world) are on the forefront of environmental change, so traditional, intergenerational knowledge adds both context and nuance to climate change discussions, research and policymaking (Curry and Lopez 2020; Rudiak-Gould 2014). Part two (and elsewhere) of this volume explores many stories of masculine, colonial exploration into Polar environments and criticises the imaginaries that these acts of heroism contributed to—but, as Kaalund (chapter "Ethnography as Racialised Womanhood in the Arctic Writings of Josephine Diebitsch-Peary"), von Spreter (chapter "Sensing Polar Ice Bodies") and others allude to, there is a lack of intersectional research that explores and expands upon diverse cultural understandings of ice and icy environments through the lens of art and aesthetics.

Part three of this book took a pop culture lens to the aesthetic of ice. The works analysed in these chapters vary from pure entertainment to intentionally educational, and range from the factual, to the fictional and stories loosely inspired by real events. As creators and audiences create, assign and rework meaning around the ice-based characters, settings and scenes portrayed in these works, our understanding of the importance and fragility of ice—and indeed those who depend on it—evolves. Perhaps these works subconsciously remind us that, as Riquet (chapter "Hard Ice, Soft Snow? Transnationalism, Spectatorship and the Arctic Sublime in Chasing Ice (2011) and Silent Snow (2011)"), Nickl (chapter "On the Visual Narratives of Ice in Popular Culture: Comics on Ice, Icy Villains and Ice Science") and others highlight, ice and human(ity) are one and the same. As we implicitly and explicitly harm ice, what, then, are we doing to ourselves? The answer lies in the power of communicating through art and aesthetic so that we make meanings together.

Looking Behind the Scenes: Into the Eco-emotional Pedagogy of Ice?

In this volume, ice appears as a stage for ecotheatre and ecodramaturgy; a stage for gender representations and personal experiences; a stage that brings communities together and builds resilience to the multiple consequences of global warming; a stage that re-energises audiences through snowmen 'clowning ice' to 'emotioneer' environmental awareness; a stage that takes up the physicality of ice to make ice regions and their loss tangible. Ice stages can be formal and informal, but are continually contested and politically charged. Some of the authors focus on active and resilient subject positions within an interconnected ice ecosystem, others concentrate on how we experience mediated ice representations. The chapters and their stories within this volume raise challenges and questions for further investigation and reflection. For example, how can the interdisciplinary nature of this research expand to incorporate Social Science techniques to provide an evidence base for 'best practice' in communicating the nature of ice and icy environments? What approaches are effective in different contexts (e.g. areas of the Global South)? The authors in this edited collection approach ice mainly from 'Western' perspectives, with an educated and industrialised background (cf. Henrich et al. 2010). Although they explicitly ask about the ignorance within our perception of ice and support the inclusion of still marginalised perspectives, including Indigenous ice knowledge, these approaches and understandings should be given even more space with regard to the participating authors in future publications.

The authors explore various strategies of how (popular) art, media and performance can draw our attention to the Anthropocene and climate change, for example by reinvigorating the fearful reverence and sensory appeal that encourages attentiveness for ice in the cultural imaginary, but also by embracing ice fabulation and enjoyment. While the role of the sublime is explicitly discussed in several chapters of this volume, the power of other communication strategies to convey a sense of ephemerality and urgency and to promote engagement with environmental issues—such as humour and visual storytelling—is worth further consideration. Humour is examined in the context of cartoons, comics and science shows, thus avoiding fear-based messaging and counteracting negative feelings audiences typically associate with climate change; in particular, laughter—as an expression of humour and a reaction to the unknown, a moment of catharsis and inversion of the sublime (Paul 1959–63) —could be further explored by future research, and this also applies to visual storytelling as a key strategy in various forms of climate-related artworks (Meretoja and Davis 2018). Not surprisingly, perhaps, there is a call to pay more attention to the power of fiction in addressing our environmental crisis, to the humanities and related fields that explore the intricacies of social processes, the nature and capacity of political change and the circulation and organisation of symbolic meaning through culture (Löschnigg and Braunecker 2019, 3–4). Fictional narratives can not only contextualise research (Lo and Huang 2021), but also generate 'emotional engagement with the characters in ways that can never be accomplished with the mere presentation of facts' (Davies et al. 2019, 9), that is the scientific presentation of environmental data. Indeed, visual and story-based communication forms have precedent in translating the sublime and the scientific together (see e.g. Finlay et al. 2021).

The climate is not just a measured result, but is closely connected to the reality of life and the mood of us humans. In times of climate change, the question of the power of images and their epistemic value is becoming increasingly important (Schneider 2018). Images that deal with climate change not only communicate knowledge, but they also give rise to a new consciousness, namely one of urgency to act and personal responsibility. When artists work with ice as a motif, metaphor or material, they do not only change the recipients' perceptions of climate change: They make insight into the accelerating and dynamic processes of climate change possible in the first place. For example, viewing melting ice in the context of an aesthetic question can convey a sense of witnessing and being physically affected. This is the cultural power that the editors attribute to ice as a medium of communication, and for whose investigation this interdisciplinary volume is intended to provide an initial point of exploration.

Bibliography

Banki, Peter. 2014. Humour as the Inverted Sublime: Jean Paul's Laughter within Limitations. *Parrhesia* 21: 58–68.

Bentz, Julia. 2020. Learning About Climate Change in, with and Through Art. *Climatic Change* 162 (3): 1595–1612. https://doi.org/10.1007/s10584-020-02804-4.

Bisbee O'Connell, Kari, Brianna Keys, Martin Storksdiek, and Mark Rosin. 2020. Context Matters: Using Art-Based Science Experiences to Broaden Participation Beyond the Choir. *International Journal of Science Education, Part B: Communication and Public Engagement* 10 (2): 166–185. https://doi.org/10.1080/21548455.2020.1727587.

Campos, Rita, and Magnólia Araújo. 2017. Traditional Artistic Expressions in Science Communication in a Globalized World: Contributions From an Exploratory Project Developed in Northeast Brazil. *Science Communication* 39 (6): 798–809. https://doi.org/10.1177/1075547017721204.

Curry, Tracie, and Ellen D.S. Lopez. 2020. Images as Information: Context-Rich Images and the Communication of Place-Based Information Through Increased Representation in Environmental Governance. *Frontiers in Communication* 5 (July): 1–15. https://doi.org/10.3389/fcomm.2020.00043.

Davies, Sarah Rachel, Megan Halpern, Maja Horst, David A. Kirby, and Bruce Lewenstein. 2019. Science stories as Culture: Experience, Identity, Narrative and Emotion in Public Communication of Science. *JCOM* 18 (05): A01:1–17. https://doi.org/10.22323/2.18050201.

Featherstone, Helen. 2014. PCST 2014. *Journal of Science Communication* 13 (03): 1–4. https://doi.org/10.22323/2.13030603.

Finlay, Summer May, Sujatha Raman, Elizabeth Rasekoala, Vanessa Mignan, Emily Dawson, Liz Neeley, and Lindy A. Orthia. 2021. From the Margins to the Mainstream: Deconstructing Science Communication as a White, Western Paradigm. *Journal of Science Communication* 20 (01): C02:1–12. https://doi.org/10.22323/2.20010302.

Henrich, Joseph, Steven J. Heine, and Ara Norenzayan. 2010. Most People Are Not WEIRD. *Nature* 466 (7302): 29. https://doi.org/10.1038/466029a.

Kuhn, Sarah, and Michael J. Muller. 1993. Participatory Design. *Communications of the ACM* 36 (4): 24–28.

Leane, Elizabeth, Carolyn Philpott, and Matt Delbridge. 2020. Performing Ice: Histories, Theories, Contexts. In *Performing Ice*, ed. Elizabeth Leane and Matt Delbridge, 1–26. London: Palgrave Macmillan.

Lesen, Amy E., Ama Rogan, and Michael J. Blum. 2016. Science Communication Through Art: Objectives, Challenges, and Outcomes. *Trends in Ecology & Evolution* 31 (9): 657–660. https://doi.org/10.1016/j.tree.2016.06.004.

Lo, Yin-Yueh, and Chun-Ju Huang. 2021. Differences in Knowledge, Uncertainty, and Social Context in Four Medical TV Series from Taiwan, Japan, South Korea and the United States. *JCOM* 20 (01): A01:1–18. https://doi.org/10.22323/2.20010201.

Löschnigg, Maria, and Melanie Braunecker. 2019. *Green Matters: Ecocultural Functions of Literature*. Netherlands: Brill.

Meretoja, Hanna, and Colin Davis. 2018. Introduction: Intersections of Storytelling and Ethics. In *Storytelling and Ethics. Literature, Visual Arts and the Power of Narrative*, ed. Hanna Meretoja and Colin Davis, 1–20. London: Routledge. https://doi.org/10.4324/9781315265018.

Quick, Kathryn S., and Martha S. Feldman. 2011. Distinguishing Participation and Inclusion. *Journal of Planning Education and Research* 31 (3): 272–290. https://doi.org/10.1177/0739456X11410979.

Rudiak-Gould, Peter. 2014. The Influence of Science Communication on Indigenous Climate Change Perception: Theoretical and Practical Implications. *Human Ecology* 42 (1): 75–86. https://doi.org/10.1007/s10745-013-9605-9.

Schneider, Birgit. 2018. *Klimabilder. Eine Genealogie globaler Bildpolitiken von Klima und Klimawandel*. Berlin: Matthes & Seitz.

Schwartz, Brian. 2014. Communicating Science through the Performing Arts. *Interdisciplinary Science Reviews* 39 (3): 275–289. https://doi.org/10.1179/0308018814Z.00000000089

Sjöberg, Ylva, Sarah Gomach, Evan Kwiatkowski, and Mathilde Mansoz. 2019. Involvement of Local Indigenous Peoples in Arctic Research—Expectations, Needs and Challenges Perceived by Early Career Researchers. *Arctic Science* 5 (1): 27–53. https://doi.org/10.1139/as-2017-0045.

Slaby, Jan, and Christian von Scheve. 2019. *Affective Societies: Key Concepts*. London: Routledge. https://doi.org/10.4324/9781351039260.

Streicher, Barbara, Kathrin Unterleitner, and Heidrun Schulze. 2014. Knowledge Rooms—Science Communication in Local, Welcoming Spaces to Foster Social Inclusion. *Journal of Science Communication* 13 (2): C03:1–5. https://doi.org/10.22323/2.13020303.

Warner, Darya. 2022. Hybrid Matters: Art and Science as a New Epistemology. *DNA and Cell Biology* 41 (1): 16–18. https://doi.org/10.1089/dna.2021.0527.

Zaelzer, Cristian. 2020. The Value in Science-Art Partnerships for Science Education and Science Communication. *eNeuro* 7 (4): ENEURO.0238-0220.2020. https://doi.org/10.1523/ENEURO.0238-20.2020.

Index[1]

[1] Note: Page numbers followed by 'n' refer to notes.

A. Hemkendreis, A.-S. Jürgens (eds.), *Communicating Ice through Popular Art and Aesthetics*, Palgrave Studies in Media and Environmental Communication,
https://doi.org/10.1007/978-3-031-39787-5

B

C

SPRINGER NATURE

GPSR Compliance

The European Union's (EU) General Product Safety Regulation (GPSR) is a set of rules that requires consumer products to be safe and our obligations to ensure this.

If you have any concerns about our products, you can contact us on ProductSafety@springernature.com

In case Publisher is established outside the EU, the EU authorized representative is:

Springer Nature Customer Service Center GmbH
Europaplatz 3
69115 Heidelberg, Germany

FSC
www.fsc.org
MIX
Papier | Fördert
gute Waldnutzung
FSC® C083411

Kolibri 360 GmbH
Ferdinand-Jühlke-Straße 7
99095 Erfurt, Deutschland
produktsicherheit@kolibri360.de